CW00722975

GAS SERVICES

GAS UTILISATION

includes Gas Pipework, Gas Supply, Pressure and Flow,Combustion and Gas Controls

Amended November 1994

Published by:
Construction Industry Training Board
Second Edition, 1991
ISBN 0 902029 87 8
© Construction Industry Training Board, 1988

CONTENTS

LIST OF TABLES

FOREWORD

The aim of all those involved in gas and gas service has always been to make certain that the appliances which the public buy are such as to combine perfect safety and the best overall efficiency in use. The maintenance of the high prestige enjoyed by the gas industry will depend on the extent to which the aim is fulfilled in practice in the future.

It follows that an understanding of what happens when gas is burned — what gas is, in what ways it can be best applied, how it can be controlled, and the common faults that occur, is essential. This publication attempts to give you that understanding, under the following headings:

Gas Pipework: This is mainly concerned with internal installation practice including the calculation of pipe sizes, design and planning, electrical bonding and testing the system. It includes a useful series of reference tables.

Gas Supply: This section is about the properties of natural gas and deals with the elementary science or technology which forms the foundation of all gas service work.

Pressure and Flow: Gas cocks and governors including adjustment and servicing.

Combustion: This section outlines the principles involved in combustion and how it works in practice. The many different sizes and shapes of flame are illustrated as well as the procedure for checking that the gas rate complies with the manufacturers recommendation.

Gas Controls: A wide range of devices have been developed to control domestic gas appliances. This section deals with these controls, their function and operation.

GAS PIPEWORK

GAS SERVICES INTO DOMESTIC PROPERTIES

All gas service pipes should be laid by the shortest practical route from the gas main to the building and, whenever practical, terminate at a British Gas meter box. The installation of such services must comply with the Gas Safety Regulations. In particular it should be noted that services cannot be routed through unventilated voids, under load-bearing footings or other foundations. Additionally, wherever a service pipe passes through a wall, the following requirements must be observed:

- Opening(s) made in the wall(s) must be sleeved to accommodate the pipe. (Suitable sleeves may be fitted by the builder or by British Gas at agreed positions.)

- Opening(s) between the sleeve and brickwork must be sealed with mortar to prevent the ingress of gas into the cavity.

- Annulus between the sleeve and pipe must be sealed. (The sealing of the sleeve will be carried out by British Gas.)

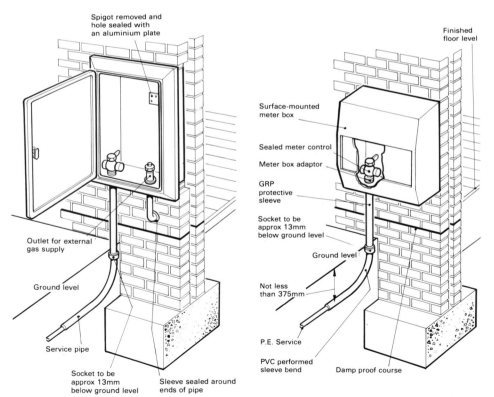

Figure 1a. Gas service termination at approved built-in meter box.

Figure 1b. Gas service termination at approved surface-mounted box.

SERVICES TO SINGLE DWELLING DOMESTIC PROPERTY

The method of termination and entry of individual services, whilst taking account of local conditions, should be at one of the meter positions described below:

- An approved built-in sunken, surface-mounted or semi-concealed meter box located on, or as near as practicable to the house wall closest to the gas main, as illustrated in Figs. 1a, 1b and Fig. 2.

- In an attached or integral garage or other suitable outbuilding, as close as possible to the external above-ground entry point of the service, as indicated in Fig. 3.

- Inside the dwelling as close as possible, (not exceeding 2 metres) to the external above-ground entry point, as indicated in Fig 4.

Where the construction of the dwelling renders it impossible to enter the building and locate the meter in accordance with any of the above alternatives, consideration will be given in full consultation with British Gas, to a meter position inside a garage or the dwelling, not exceeding 2 metres from the position of a below-ground entry.

No further builders work is required for the provision of the gas service pipe which will be directly laid by British Gas.

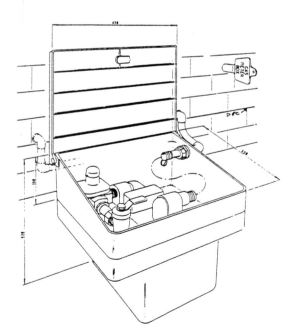

Figure 2. Gas service termination at approved semi-concealed meter box.

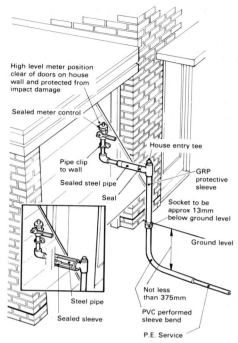

Figure 3. Gas service termination in attached or integral garage.

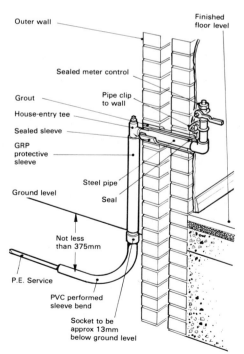

Figure 4. Gas service termination inside a dwelling.

3

A. TRADITIONAL BRICK

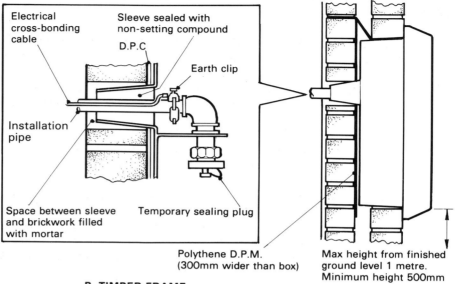

Electrical cross-bonding cable

Sleeve sealed with non-setting compound

D.P.C

Earth clip

Installation pipe

Space between sleeve and brickwork filled with mortar

Temporary sealing plug

Polythene D.P.M. (300mm wider than box)

Max height from finished ground level 1 metre. Minimum height 500mm

B. TIMBER FRAME

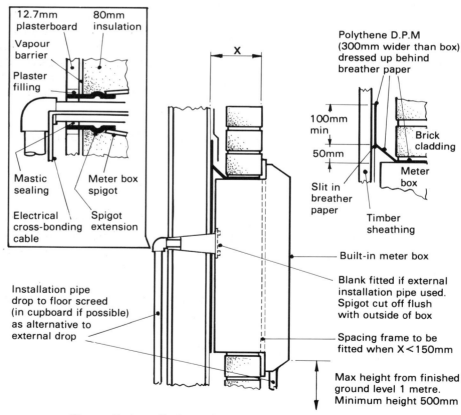

12.7mm plasterboard

80mm insulation

Vapour barrier

Plaster filling

X

Polythene D.P.M (300mm wider than box) dressed up behind breather paper

100mm min

50mm

Brick cladding

Meter box

Mastic sealing

Meter box spigot

Slit in breather paper

Timber sheathing

Electrical cross-bonding cable

Spigot extension

Built-in meter box

Blank fitted if external installation pipe used. Spigot cut off flush with outside of box

Installation pipe drop to floor screed (in cupboard if possible) as alternative to external drop

Spacing frame to be fitted when X<150mm

Max height from finished ground level 1 metre. Minimum height 500mm

Figure 5. Installation of approved built-in meter box.

4

SERVICES TO MULTI-STOREY AND MULTI-DWELLING PROPERTIES

Supplies to multi-storey and multi-dwelling properties, services should terminate at one of the meter positions described below, taking account of local conditions.

- In approved, readily accessible, built-in sunken or surface-mounted meter boxes with external service risers located on, or as near as practicable to the house wall closest to the gas main.

- Inside the dwelling, as close as possible but not exceeding 2 metres from the position of a readily accessible external riser.

- Inside the dwelling as close as possible but not exceeding 2 metres from the position of an internal service riser within a protected shaft. The riser should be sited to minimise the length of any internal pipework or laterals and should be as close as possible, but not exceeding 2 metres, from the entry point of the service, as indicated in Figure 6.

Surface-Mounted Meter Box

The surface-mounted box has been developed as an alternative to the built in sunken box and is particularly useful in older houses when conversions or modernisation programmes are carried out. Of lightweight construction, the surface-mounted box is manufactured in three parts — backplate, cover and door. The complete meter installation can be assembled on the backplate after it has been screwed to the wall, finally adding the cover and door.

The dimensions of the complete assembly are approx. 506mm wide by 230mm deep.

PROVISION OF INTERNAL SHAFTS

Attention is drawn to the provisions of the Building Regulations for services routed inside multi-storey buildings. Such services can only be installed inside the building in specially protected shafts. Located adjacent to an external wall, the protected shaft must be of one of the following types:

A continuous shaft ventilated to the outside of the building at the top and bottom levels of the shaft. At each point where a lateral service branches from the main riser through the wall of the shaft, a fire-stopped sleeve must be provided (see Figure 7).

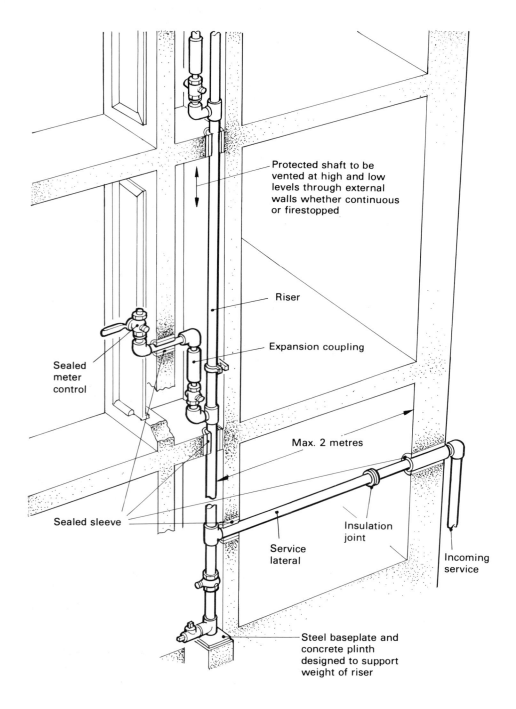

Protected shaft to be vented at high and low levels through external walls whether continuous or firestopped

Riser

Expansion coupling

Sealed meter control

Max. 2 metres

Sealed sleeve

Insulation joint

Incoming service

Service lateral

Steel baseplate and concrete plinth designed to support weight of riser

Figure 6. Gas service entry and termination from a vertical riser in mutli-storey and multi-dwelling property.

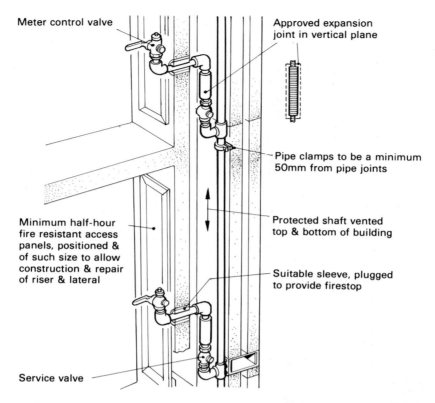

Figure 7. Gas service routed in protected, continuous and ventilated internal shaft.

Service Laterals

All service laterals must incorporate a valve adjacent to the riser and be constructed with flexible connections to allow for differential thermal expansion.

Internal Risers

The service riser must be adequately supported at its point of contact above the base to prevent settlement. To facilitate this, a steel base-plate and concrete plinth must be provided to transmit the load. For maintenance purposes, access to the base of the riser must be provided.

The riser must not deviate in direction from the vertical plane and bracket supports must be provided at specified intervals. Account should be taken at the structural-design stage of the shaft to allow for the transmission of these loads.

Where there are more than five storeys, the riser will be welded and extra allowances should be made for this aspect of contruction.

Cocks and valves

A cock or valve (designated the emergency control) should always be fitted:

(1) At the inlet of the primary meter (see BS 6400); or

(2) to the installation pipe where it enters the building where the meter is sited 6 m or further away from the building, or;

(3) inside individual flats served by a large single or a multiple meter installation located in a remote or communal area.

NOTE:

(A) Where this valve is fitted at the inlet of a primary meter and is not in an accessible position, has not handle or is difficult to operate the gas supplier should be advised accordingly.

(B) Meters may be located in remote areas.

(C) Meters may be grouped together in a hallway or on the outer wall of a block of flats.

Plug cocks must comply with BS 1552.

A test point shall be located downstream of an emergency control for soundness testing purposes.

The emergency control shall:

(a) be labelled or otherwise marked to show the open and closed positions;

(b) be fitted in an accessible position;

(c) be easy to operate;

(d) be fitted with a suitable handle or lever which is securely attached.

(e) where the emergency control does not form part of the primary meter installation, a permanent notice bearing the words 'Gas Emergency Control' shall be fixed in a prominent position or near to the valve, indicating to the consumer(s) the action to be taken in the event of an escape of gas, the name and telephone number of the gas supplier, and the date on which the notice was first displayed.

Suitable notices are generally available from the supplier of gas.

Where the installation pipe leaves the building to supply remote appliances (i.e. greenhouse heaters, barbecues and lighting) a cock or valve shall be fitted either internally or externally at the point of exit.

8

INSTALLATION PRACTICE

The British Standard 'Glossary of Terms used in the Gas Industry' does not recognise the term 'internal installation'. It refers instead to 'installation pipes'. However, the term 'internal installation' is universally understood to be an installation within premises, including the meter, governor and any other control devices between the meter control and the points to which the appliances are connected.

This is (at the moment) at variance with the continental definition which has internal installation starting at the meter outlet and including the appliances. Although the British definition includes the meter, this section deals with the pipework.

The installation is also sometimes called the 'carcass' or house carcass. This term is depreciated by the 'Glossary' but it has persisted in use and is specifically applied to the pipework installed in new houses whilst they are under construction. 'Carcassing' is generally understood to mean working on a building site, installing pipework before the floors are laid or the walls plastered.

The installation of pipes in new buildings should ideally be planned at the drawing board stage, when the meter and appliance positions and the pipe runs may be indicated on the plans. The installation is phased in with the work of the other building crafts so that all the building engineering services can be installed immediately the shell of the building and the floor joists are completed.

Installations should be carried out in accordance with BS Specification 6891:1988 and must also comply with the Gas Safety Regulations. These requirements are embodied in the practices recommended in this section.

Both the person carrying out the installation and the employer are equally responsible for ensuring that the Gas Safety Regulations are satisfied. Any person offending against the regulations can be prosecuted.

BS 6891: 1988

The British Standard specifies the design, materials and methods of installing gas pipework, in sizes not exceeding 28 mm (type R1 of BS 21) for 2nd family gas installation of a domestic type.

It applies to low pressure installations having a maximum pressure of 75 mbar (30 in wg).

Service pipes are covered by the *Institution of Gas Engineers* Recommendations IGE/TD/4: Edition 2.

L.P.G. installations are covered by B.S. 5482 Parts I & II.

Definitions

The definitions given in BS 1179: 1967 and in BS 1179: Part 6: 1980 apply, and include the following:

Installation pipes. That, or any part of the installation pipework from the first fitting or joint after the primary meter outlet union connection to points at which appliances are to be connected.

Duct. A purpose designed enclosure to contain gas pipes and having a cross-sectional area greater than 10,000 mm².

Emergency control. A cock or valve for shutting off the supply of gas in an emergency.

Meter control. The valve fitted upstream of, and adjacent to, a meter to shut off the supply of gas to it.

Primary meter. A meter connected to the service pipe, the index reading of which constitutes the basis of charge for all gas used on the premises.

Service pipe. A pipe connected to a main to provide a supply of gas to one or more consumers and terminating at and including the primary meter control(s).

Sleeve. A duct, tube or pipe embedded in the structure for the reception of an installation pipe.

Installation Practice (cont'd)

Design and Planning

At the initial stages of building design and planning the interested parties should verify that the installation pipes will be adequate for both immediate and probable future requirements.

The necessary information concerning the routing of installation pipes and positions of controls and appliance points should be made available by means of drawings, specifications and consultations.

Any installation pipe which is fitted as the erection of a building progresses, and which will subsequently be inaccessible, must be tested for soundness before being buried, covered or wrapped.

Methods for the calculation of pipe sizes

When designing the installation, the size of all installation pipes will be determined by the maximum gas rate of the appliances to be connected, allowance being made for the maximum demand likely to occur at any time.

The possibility of future extensions must also be considered, particularly if the pipes are to be buried.

The pressure drop between the outlet of the meter and appliances connected must not exceed 1 mbar under maximum flow condition (see Tables 1 and 2).

Table 1. Discharge in a straight horizontal steel pipe with 1.0 mbar (0.4 in w.g.) differential pressure between the ends, for gas of relative density 0.6 (air = 1)

(a) Piping in accordance with table 4 (medium) of BS 1387 : 1985

Nominal size		Length of pipe (m)															
		3		6		9		12		15		20		25		30	
		Discharge															
mm	in	m³/h	ft³/h	m³/h	ft³/h	m³/h	ft³/h	m³/h	ft³/h	m³/h	ft³/h	m³/h	ft³/h	m³/h	ft³/h	m³/h	ft³/h
6	⅛	0.29	10	0.14	4.9	0.09	3.2	0.07	2.5	0.05	1.8	–	–	–	–	–	–
8	¼	0.8	28	0.53	19	0.49	17	0.36	13	0.29	10	0.22	7.8	0.17	6.0	0.14	4.9
10	⅜	2.1	73	1.4	49	1.1	39	0.93	33	0.81	30	0.70	29	0.69	24	0.57	20
15	½	4.3	150	2.9	100	2.3	82	2.0	70	1.7	61	1.5	52	1.4	46	1.3	44
20	¾	9.7	340	6.6	230	5.3	190	4.5	160	3.9	140	3.3	120	2.9	100	2.6	93
25	1	18	650	12	440	10	350	8.5	300	7.5	260	6.3	220	5.6	200	5.0	180

NOTE. When using this table to estimate the gas flow rate in pipework of a known length, this length should be increased by 0.5m (2ft) for each elbow and tee fitted, and by 0.3m (1ft) for each 90° bend fitted.

(b) Piping in accordance with table 5 (heavy) of BS 1387 : 1985

Nominal size		Length of pipe (m)															
		3		6		9		12		15		20		25		30	
		Discharge															
mm	in	m³/h	ft³/h	m³/h	ft³/h	m³/h	ft³/h	m³/h	ft³/h	m³/h	ft³/h	m³/h	ft³/h	m³/h	ft³/h	m³/h	ft³/h
6	⅛	0.11	3.9	0.055	2.0	0.037	1.3	0.028	0.98	0.022	0.78	0.017	0.59	0.013	0.47	0.011	0.39
8	¼	0.56	20	0.43	15	0.28	10	0.21	7.5	0.17	6.0	0.13	4.5	0.10	3.6	0.085	3.0
10	⅜	1.6	56	1.1	38	0.85	30	0.79	28	0.77	27	0.59	21	0.47	17	0.40	14
15	½	3.5	120	2.3	83	1.9	68	1.6	56	1.4	49	1.2	41	1.0	36	0.92	32
20	¾	8.3	290	5.6	200	4.5	160	3.8	130	3.3	120	2.8	100	2.5	88	2.2	79
25	1	15	550	11	370	8.4	300	7.1	250	6.3	220	5.3	190	4.7	170	4.2	150

NOTE. When using this table to estimate the gas flow rate in pipework of a known length, this length should be increased by 0.5m (2ft) for each elbow and tee fitted, and by 0.3m (1ft) for each 90° bend fitted.

Table 2. Discharge in a straight horizontal copper tube with 1.0 mbar (0.4 in w.g.) differential pressure between the ends, for gas of relative density 0.6 (air = 1)

(a) Piping in accordance with table X of BS 2871 : Part 1 : 1971

Size of tube (mm)	Length of pipe (m) — Discharge															
	3		**6**		**9**		**12**		**15**		**20**		**25**		**30**	
	m³/h	ft³/h	m³/h	ft³/h	m³/h	ft³/h	m³/h	ft³/h	m³/h	ft³/h	m³/h	ft³/h	m³/h	ft³/h	m³/h	ft³/h
10	0.86	30	0.57	20	0.50	18	0.37	13	0.30	11	0.22	7.8	0.18	6.4	0.15	5.3
12	1.5	54	1.0	36	0.85	30	0.82	29	0.69	24	0.52	18	0.41	14	0.34	12
15	2.9	100	1.9	69	1.5	54	1.3	45	1.1	40	0.95	34	0.92	32	0.88	31
22	8.7	310	5.8	210	4.6	160	3.9	140	3.4	120	2.9	100	2.5	89	2.3	80
28	18	630	12	420	9.4	330	8.0	280	7.0	250	5.9	210	5.2	180	4.7	170

NOTE. When using this table to estimate the gas flow rate in pipework of a known length, this length should be increased by 0.5m (2ft) for each elbow and tee fitted, and by 0.3m (1ft) for each 90° bend fitted.

(b) Piping in accordance with table Y of BS 2871 : Part 1 : 1971

Size of tube (mm)	Length of pipe (m) — Discharge															
	3		**6**		**9**		**12**		**15**		**20**		**25**		**30**	
	m³/h	ft³/h	m³/h	ft³/h	m³/h	ft³/h	m³/h	ft³/h	m³/h	ft³/h	m³/h	ft³/h	m³/h	ft³/h	m³/h	ft³/h
6	0.082	2.9	0.040	1.4	0.028	0.98	0.021	0.73	0.017	0.59	0.011	0.39	0.010	0.35	0.0083	0.29
8	0.40	14	0.20	7.0	0.13	4.7	0.10	3.6	0.081	2.8	0.054	1.9	0.047	1.7	0.039	1.4
10	0.74	26	0.60	21	0.40	14	0.32	11	0.24	8.6	0.17	5.8	0.15	5.2	0.12	4.3
12	1.4	48	0.91	32	0.74	26	0.71	25	0.57	20	0.39	14	0.36	13	0.28	10
15	2.5	90	1.7	60	1.3	47	1.1	40	0.98	35	0.96	34	0.86	30	0.70	25
22	7.9	280	5.3	190	4.2	150	3.5	120	3.1	110	2.6	92	2.3	81	2.1	73
28	16	580	11	390	8.8	310	7.4	260	6.5	230	5.5	190	4.8	170	4.3	150

NOTE. When using this table to estimate the gas flow rate in pipework of a known length, this length should be increased by 0.5m (2ft) for each elbow and tee fitted, and by 0.3m (1ft) for each 90° bend fitted.

Installation Practice (cont'd)

Table 3 The effects of elbows, tees or bends inserted in a run of pipe (expressed as the approximate additional lengths to be allowed)

Nominal size				Approximate additional lengths to be allowed					
Cast iron or mild steel		Stainless steel or copper		Elbows		Tees		90° bends	
mm	(in)	mm	(in)	m	(ft)	m	(ft)	m	(ft)
Up to 25	(1)	Up to 28	(1)	0.5	(2)	0.5	(2)	0.3	(1)
32 to 40	(1¼ to 1½)	35 to 42	(1¼ to 1½)	1.0	(3)	1.0	(3)	0.3	(1)
50	(2)	54	(2)	1.5	(5)	1.5	(5)	0.5	(2)
80	(3)	76.1	(3)	2.5	(8)	2.5	(8)	1.0	(3)

Example (see Diagram opposite)

Assume a domestic installation supplying a cooker (1 m^3/h of gas), a boiler (1.6 m^3h) and a gas fire (0.5 m^3/h) with dimensions as shown below using copper pipe of size and length as given in Table 2:

NOTE. The maximum permitted pressure loss from the meter A to any point of use is 1 mbar. (0.4" WG) under maximum flow conditions.

Section A – B

The flow is 3.1 m/h and the section includes two elbows and a tee. The corrected length is therefore 3 m + (3 x 0.5 m) or 4.5 m (from table 3).

Assume the use of 22 mm copper pipe. Table 2(a) shows that 3.4 m/h of gas in 15 m of 22 mm pipe gives a loss of 1 mbar.

4.5 m (4.5 m ÷ 15 m) of 22 mm pipe will therefore have a pressure loss of 0.3 mbar.

cont'd on page 16

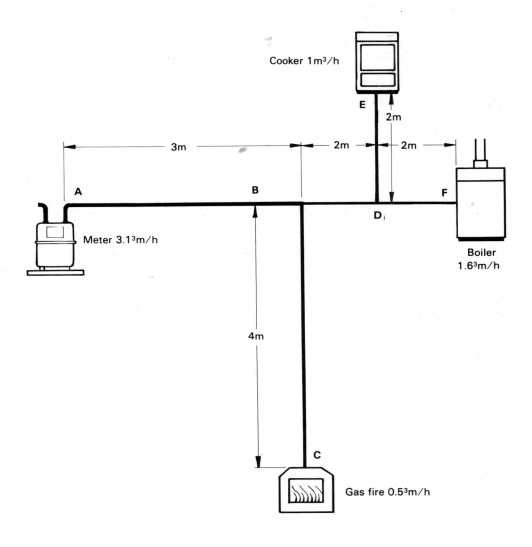

Example - effect of elbows, tees and bends

Section B — C

The flow is 0.5 m/h and the section includes three elbows.

The corrected length is therefore 4 m + (3 × 0.5 m) or 5.5 m.

Note: For natural gas of 38.16 MJ/m as supplied by British Gas plc 1 kW of heat input is equivalent to 0.094 m/h and 1 m/h is equivalent to 10.6 kW of heat input.

Assume the use of 15 mm pipe. Table 2(a) shows that 0.88 m/h of gas in 30 m of pipe will give a pressure loss of 1 mbar. Therefore 5.5 m (5.5 ÷ 30 m) of 15 mm pipe will have a pressure loss of approximately 0.2 mbar.

The total pressure loss from A to C is therefore 0.3 mbar + 0.2 mbar or 0.5 mbar which is acceptable but indicates that a smaller pipe might be used.

Assume that 12 mm pipe were used instead of the 15mm. Table 2(a) shows that 0.52 m/h will pass through 20 m of pipe with a pressure loss of of 1 mbar. Therefore through 5.5 m of pipe the pressure loss will be approximately 0.3 mbar.

This gives a total pressure loss to C of 0.6 mbar, which is also acceptable being below 1 mbar.

Section B — D

The flow is 2.6 m/h and the section includes three elbows and a tee. The corrected length is therefore 2 m + (4 × 0.5 m) or 4 m.

As with previous examples, table 2(a) shows that 2.9 m/h will pass through 20 m of 22 m pipe with a pressure loss of 1 mbar. Therefore through 4 m of pipe the loss will be approximately 0.3 mbar.

The total pressure loss from A to D is therefore 0.3 mbar + 0.2 mbar or 0.5 mbar.

Section D — E

The flow is 1 m/h and the section includes two elbows. The corrected length is therefore 2 m + (2 × 0.5 m) or 3 m.

As with previous examples, table 2(a) shows that 15 mm pipe will produce a pressure loss of 0.2 mbar giving a total pressure loss to E of 0.3 mbar + 0.2 mbar + 0.2 mbar or 0.7 mbar, which is acceptable.

Section D — F

The flow is 1.6 m/h and the section includes two elbows. The corrected length is therefore 2 m + (2 × 0.5 m) or 3 m.

As with previous examples, table 2(a) shows that 15mm pipe will produce a pressure loss of 0.5 mbar giving a total pressure loss to F of 0.3 mbar + 0.2 mbar + 0.5 mbar or 1.0 mbar, which is acceptable.

Installation Practice (cont'd)

Resistance

Installation pipes must not be restricted by kinks, burrs, foreign matter etc.

Fittings must be kept to a minimum. Bends should be used in preference to elbows.

Materials

Consideration should be given to the strength, appearance and cost of materials. The need for protection against corrosion, must also be considered.

All pipes and fittings must comply with the relevant British Standards.

Steel pipes and fittings should comply with BS 1387 medium or heavy grade, BS 3601 and BS 3604.

Stainless steel pipes should comply with BS 3605 or BS 4127 and be jointed with compression fittings complying with BS 864: Part 2.

Malleable iron fittings should comply with BS 143 and BS 1256.

Copper tube should comply with table X or Y BS 2871: Capillary and compression fittings with BS 864: Part 2

Compression fittings must not be buried in the structure, below ground or beneath floorboards.

Plastic/Polyethylene pipes and fittings should only be used for exterior pipework where buried or otherwise protected against light and mechanical damage.

Jointing All **compounds** and tapes must comply with BS 5292.

Capilliary joints Fittings must comply with BS 864: Part 2. Finished joints must be visually examined to confirm that the solder has run.

The flux used should be non-corrosive or aggressive after the soldering operation. Self cleaning fluxes may fall into this category.

Copper pipes may be jointed by the use of purpose-made tools that form the pipe ends in accordance with BS 864: Part 2.

Installation Practice (cont'd)

Compression Fittings must comply with BS 864: Part 2, and must be readily accessible.

Unions shall only be located in accessible positions and be of the ground face or compression type.

Screwed Fittings used for threaded joints shall comply with BS 143 & 1256. Threads on long screws must be parallel. Threads on all other fittings except those to BS 2779, should be tapered in accordance with BS 21.

The threads must be cleaned before use.

Hemp must not be used on a threaded joint except when in conjunction with thread sealing compounds for long screw back-nut seals.

PTFE tape shall be wound with a 50% overlap starting from the thread runout in a direction counter to the thread form (see Fig. 8).

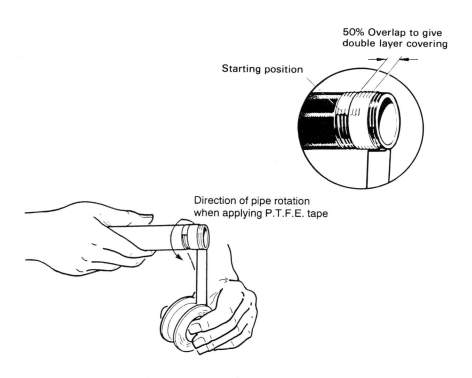

Figure 8. Thread wrapping method for PTFE tape

While work is in progress, care must be taken to prevent the ingress of dirt, water, etc. into the pipe.

Where work is carried out on pipes already connected to a meter either:

(a) the meter must be temporarily disconnected with both the open ends of the pipework sealed and dust caps fitted to the meter; or

(b) all open ends of the pipe must be plugged or capped before the work is left unattended.

Where working with a naked flame where the pipework contains or has contained gas, the supply must be isolated and disconnected. Open ends connected to the gas supply and of any gas meter shall be plugged or capped.

Pipes laid in wooden joisted floors

Pipes which are installed between joists in ceiling or roof spaces should be properly supported.

Pipes which are laid across joists in the ceiling or roof space should be located in purpose-made notches or circular holes.

Joists of less than 100 mm depth must not be notched. The depth of notch should be sufficient to accommodate only the pipe or fittings and not exceed 15% (one-sixth) of the joist depth. The notch should be U-shaped and located not further than one-quarter of the span from the end support and of minimum width.

Care should be taken when repairing floor boards to prevent damage to the pipes by nails or screws. The boards should be appropriately marked.

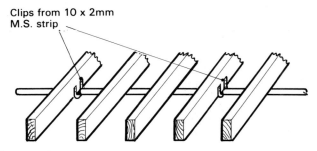

Clips from 10 x 2mm
M.S. strip

Suggested method of securing pipes under ground floor joists where working-room is restricted

Figure 9. Pipes secured under joists

Installation Practice (cont'd)

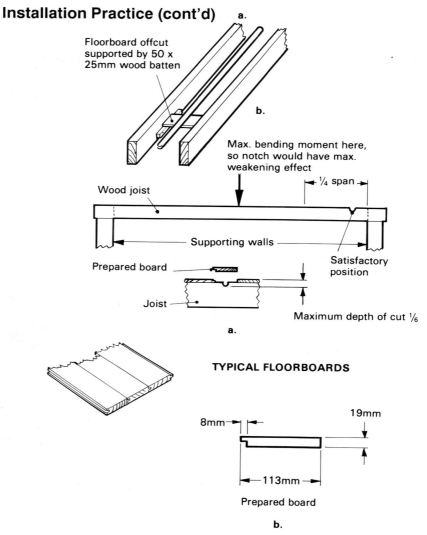

a.

Floorboard offcut supported by 50 x 25mm wood batten

b.

Max. bending moment here, so notch would have max. weakening effect

Wood joist

¼ span

Supporting walls

Satisfactory position

Prepared board

Joist

Maximum depth of cut ⅙

a.

TYPICAL FLOORBOARDS

8mm

19mm

113mm

Prepared board

b.

Figure 10. Pipes crossing on joists: (a) notch in joist; (b) covering floor board

Solid Floors

Wherever possible, the pipes should be laid on the base concrete. When laid in concrete floors, the pipe must be protected against failure caused by movement. Joints should be kept to a minimum, and compression fittings must not be used.

Four suitable methods of protection are: —

- Continuous fully annealed factory sheathed soft copper tube passed through a larger plastic tube previously set into the floor slab and/or base hard core.

- Steel or copper pipe laid on top of the base concrete and covered by a screed.

- Steel or copper laid into a preformed duct with a suitable protective covering.

- Steel or copper pipe fitted with a soft covering material which should be thick enough to accommodate movement and resilient enough to support the concrete cover during setting.

Stainless steel pipework must not be buried in solid floors.

Galvanized or painted pipes should not be buried without additional protection.

Pipes to be buried in magnesium-oxy-chloride or magnesite flooring, should be of copper with a factory bonded sheath and jointed with copper capillary fittings. Bends and joints must be protected by wrapping with a suitable plastic tape.

External buried pipes should be of polyethylene, factory sheathed copper or factory wrapped steel. All metallic joints must be fully wrapped.

Pipes passing through solid floors should take the shortest practicable route and must be sleeved.

NOTE:

The correct procedures for laying buried pipes is given in Institution of Gas Engineers IGE TD/4: Edition 2. Pipes in open soil or below vehicular traffic areas must have a cover of at least 375 mm.

Pipes in or below concrete with only pedestrian traffic should have a cover of at least 40 mm.

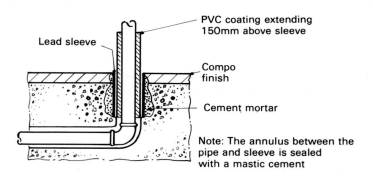

Figure 11. Pipe sleeve in composition floor

Installation Practice (cont'd)

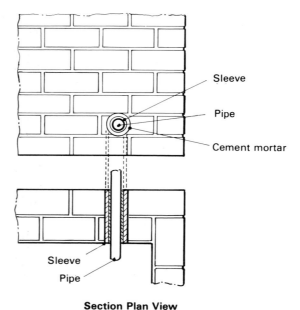

Section Plan View
Figure 12. Pipe sleeve

Pipes in walls

Pipes passing through walls must be sleeved (see Fig. 12)

Vertical pipe runs Where suitable, vertical pipes should be placed in ducts which are accessible. If this method is impractical and provided the walls are thick enough, pipes may be placed in pipe chases.

Cavity walls Pipes should not be placed within cavity walls. Pipes passing through cavity walls should take the shortest practicable route and must be sleeved.

Dry lined walls Installation pipes installed within dry lined or timber construction walls shall be:

- Run within a purpose designed channel or duct constructed and sealed as to prevent the transfer of gas into the cavity wall:
- Be suitably secured;
- Have the minimum number of joints;
- Properly protected against corrosion;
- Be protected along its full length by a metal plate or cover (for copper pipe);
- Have adequate ventilation openings at high level to allow any escape of gas to dissipate into the room.

(See Figs. 13, 14)

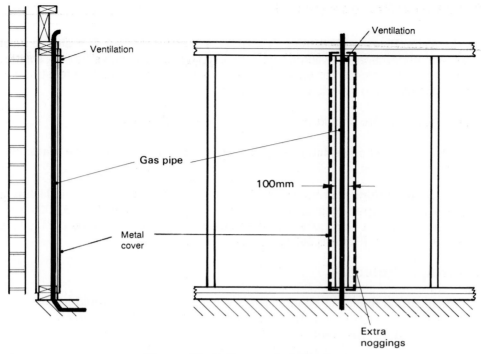

Figure 13. Full story height riser

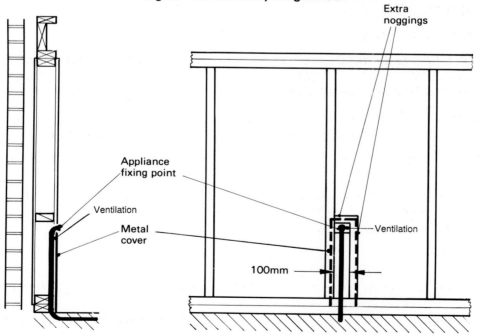

Figure 14. Appliance fixing point riser

Note

The ventilation shown above may not be required. See "Pipes in ducts" page 24.

23

Installation Practice (cont'd)

Solid Walls. Pipes passing through solid walls must be sleeved. The **sleeve** must be of a material capable of containing or distributing gas, e.g. polyethylene, copper, steel, polyvinyl chloride, or other suitable material and be sealed, at least at one end to the pipe with a flexible fire resistant mastic.

Compression joints should not be located within a sleeve.

The sleeve should be sealed to the building structure with a suitable materials, e.g. cement mortar.

Pipes in ducts Vertical or horizontal ducts containing pipes must be ventilated at the top with an opening grille having a free area of not less than 3,000 mm^2. Ventilation may not be required when the cross sectional area of the duct is less than 10,000 mm^2. (Ref. BS 6891).

Corrosion Protection

A pipe or fitting must not be installed in a position where it would be exposed to corrosion. Pipes or fittings that are at risk should either be manufactured from materials that are resistant to corrosion or be suitably protected.

Where wrapping tape is to be applied, the pipe should be clean, dry and prepared in accordance with manufacturer's instructions. At least a 50% overlap should be used.

The wrapping should be coloured yellow ochre in accordance with BS 1710.

Pipes in fireplace openings should be factory sheathed and/or wrapped on site to prevent corrosion by soot and debris.

Pipework must be tested for soundness before any additional protection against corrosion is applied.

Supports and fixings

Installation pipes should be adequately supported (see tables below).

The support used should be designed to prevent the pipe coming into contact with surfaces of the structure which are likely to cause corrosion. Suitable supports include those made from metal (see BS 1464 and BS 3974: Part 1) and plastic.

Maximum interval between pipe supports (cast iron, mild steel and stainless steel pipes)

Nominal size			Interval for vertical runs		Interval for horizontal runs	
Cast iron, mild steel	Stainless steel					
mm	mm	in	m	ft	m	ft
15	15	½	2.5	8	2.0	6
20	22	¾	3.0	10	2.5	8
25	28	1	3.0	10	2.5	8
32	35	1 ¼	3.0	10	2.7	9
40	42	1 ½	3.5	12	3.0	10
50	54	2	3.5	12	3.0	10
80	76.1	3	3.5	12	3.0	10
100	108	4	3.5	12	3.0	10

Table 4.

Maximum interval between pipe supports (light gauge copper pipes)

Nominal size		Interval for vertical runs		Interval for horizontal runs	
mm	in	m	ft	m	ft
up to 15	up to ½	2.0	6	1.2	4
22	¾	2.5	8	1.8	6
28	1	2.5	8	1.8	6
35	1 ¼	3.0	10	2.5	8
42	1 ½	3.0	10	2.5	8
54	2	3.0	10	2.7	9
—	2 ½	3.5	12	3.0	10
76.1	3	3.5	12	3.0	10
108	4	3.5	12	3.0	10

Table 5.

External pipes

The pipe must be located where it is not liable to mechanical damage. Pipes should be protected against corrosion by wrapping, painting or by pipe material selection.

Pipes in relation to other services

Pipes must be so located (or suitably electrically insulated) so that they do not come into contact with other metallic fitments which could give rise to electrolytic corrosion.

Electrical services Care should be taken not to cause damage to electrical conductors whilst working.

Pipes must not be buried in floors where electrical under-floor heating is installed. Pipes which are not protected by electrical insulating material, should be spaced as follows: —

- at least 150 mm (6") away from electricity metres and associated control devices.

- at least 25 mm (1") away from electricity cables and other metallic services.

Cross bonding. Where electrical cross bonding is necessary, a clamp should be used to make a connection to the outlet side of the primary gas meter. The connection must be as near as practicable and preferably not further than 600 mm (2 ft) of pipework from the meter outlet. Conductors connected to the earth terminal should be of a size as laid down in the I.E.E. Regulations.

Temporary Continuity bonds

Where any work involves the connection or disconnection of pipework, a temporary continuity bond must be fixed. See below.

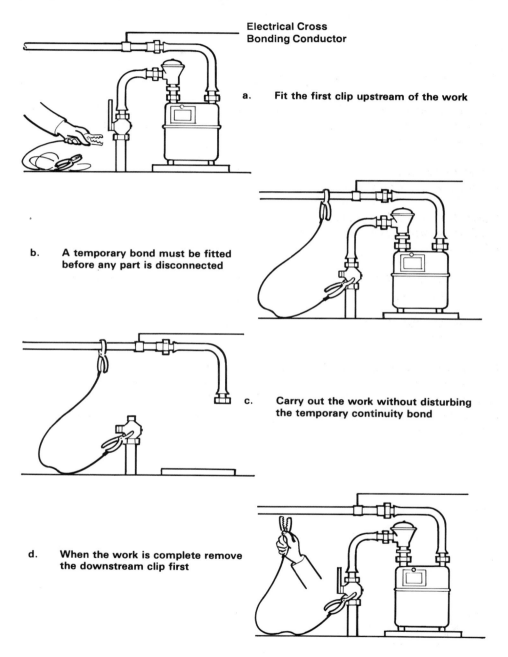

Electrical Cross Bonding Conductor

a.   **Fit the first clip upstream of the work**

b.   **A temporary bond must be fitted before any part is disconnected**

c.   **Carry out the work without disturbing the temporary continuity bond**

d.   **When the work is complete remove the downstream clip first**

Figure 15a-d. Temporary continuity bonding

TESTING THE SYSTEM

Testing New Pipework

Installations not connected to the gas supply should be tested for soundness prior to the installation of the primary meter. A test tee is a valuable aid when pressurising pipework for the soundness test.

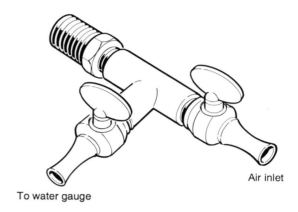

To water gauge

Air inlet

Figure 16. A typical test tee

Where appliances are already connected to a new installation, they must be isolated prior to the soundness test.

When tested there shall be no pressure loss. If the installation fails the test the fault(s) must be traced and and rectified.

When the installation has passed the test, it should either be plugged or capped at all outlets and inlets or a meter fitted. The installation should be purged.

Existing installations

When tested in accordance with procedure outlined below.

(a) Installations not recently connected to a gas meter and installations without appliances connected must not indicate any pressure loss during the 2 minute test period;

(b) Installations with appliances connected must not exceed the pressure loss values given in Table 6 below during the 2 minute test period. There must be no smell of gas.

Maximum permissible pressure drop during 2 min test period for existing installations

The permissible pressure drops given relate to average lengths of pipework in domestic installations. Where the installation pipe is connected to pipe sizes larger than 28 mm (R1) reference should be made to CITB Study Notes ME 211/1 Large Installation Pipework.

Where the installation fails the test, the fault must be traced and rectified or made safe.

Extensions to existing installation already connected to a gas supply

The existing system should be tested prior to commencing work as detailed above. Any pressure loss in the 2 minute period must be recorded. If the pressure loss exceeds the values given in the Table (below) the fault must be traced and rectified or otherwise made safe.

Table 6. Maximum permissible pressure loss for installations with appliances fitted

Meter			Pressure drop	
Designation	Capacity/Rev.			
	ft^3/r	dm^3/r	mbar	in w.g.
U6	0.07	2.0	4.0	1.6
P1	0.1	2.8	2.5	1.0
P2	0.2	5.7	1.5	0.6
P4	0.4	11.4	0.5	0.2

NOTE. This table is based on an average length of pipework in domestic installations. Where the existing installation is unusually long, the permissable pressure drop will be smaller, and reference should be made to the local gas undertaking.

On completion of an extension to the system, the whole system should be tested. If the pressure loss is no greater than that previously recorded, the system may be considered sound, provided there is no smell of gas.

Where the extension is a major addition to the installation, it should be tested separately as if it were new pipework before connecting it to the installation. The whole system should then be tested as detailed above.

Soundness test procedure

Where appliances are connected, check that they are isolated or that all operating taps and pilots are turned off.

Connect a 'U' tube pressure gauge to a suitable pressure test point on the installation.

Raise the pressure in the pipework to 20 mbar and turn off the gas supply making sure that the test pressure does not exceed 25 mbar.

Allow 1 minute for temperature stabilisation. If a pressure rise is observed either the temperature of the system is rising, in which case (a) the temperature has not stabilized adequately; or (b) the means of isolation from the pressurizing source is leaking.

If the means of isolation is leaking, the leak should be rectified. In the case of a leaking meter control, this must be reported to the **fuel supplier** immediately for rectification.

Record any pressure loss in the next 2 min and check that there is no smell of gas.

On completion of a satisfactory test either plug or seal to open end of the pipework. Where a meter is available purge pipework and test all purge and pressure points using leak detection fluid.

Purging

Every installation should be purged after passing the soundness test and before being connected to the gas supply.

During the purging operations, care must be taken to see that purged gas does not accumulate in any confined space. Steps should be taken within the vicinity of the purge point to ensure good ventilation, to prevent inadvertent operation of any electric switches or appliances and to prohibit smoking or naked lights.

Purging is effected by passing a volume of gas not less than five times the badged capacity per revolution of the meter mechanism. Pipes should be purged commencing at the point(s) furthest from the meter.

THE "U" GAS PRESSURE GAUGE

The "U" pressure gauge most frequently used by gas installers now incorporates a dual scale utilizing 'milliars' as well as 'inches' and is available in 30 mbar/12 ins W.G. size complete with protective cover which also acts as a stand. The scale can also be adjusted slightly to facilitate zeroing.

Manufacturers are increasingly listing the pressure settings of their appliances in 'millibars' as well as 'inches' and in these circumstances the gauge should be at hand at all times, not only for pressure testing but for soundness testing and checking pressure loss through meters and pipework.

Errors in reading the gauge can be brought about by wrongly positioning the eyes in relation to the water level and the scale.

When taking readings, the eyes, scale and mensicus (water level) must be at the same level. The gauge must also be vertical.

1. Using the "U" Gauge to take Pressure Readings

1.1 Check that the gauge is in good condition and that the glass is clean.

1.2 Fill the gauge to the zero position with clean water and ensure that there are no air bubbles and that the gas and air limbs are unobstructed. The scale may be adjusted slightly to obtain correct zero.

1.3 The gauge must be supported firmly in a vertical position by using the protective case cover as a stand.

1.4 Connect the gauge to the pressure point on the appliance using rubber tubing of sound condition. Connect the tube to either entry on the "U" gauge depending on whether the "inches" or "millibars" scale is to be used.

1.5 Admit the gas slowly to prevent a surge in pressure which will blow the water out of the gauge.

1.6 Read from the correct level to avoid errors in reading, and ensure the water level reads the same in *both limbs* of the gauge.

1.7 In all cases the vertical distance between the zero and liquid levels should be expressed in either "inches WG" or "millibars".

2. Procedure for Dealing with Gas Escapes

2.1 The treatment of gas escapes must at all times be given priority. The degree of smell is no indication of the seriousness of a gas escape and all complaints of a smell of gas must be treated with urgency.

2.2 There are two points of possible gas escapes, internal, ie from the house carcass or appliances and external, ie from the mains or service pipes within the road.

The former are under the control of the customer control valve adjacent to the gas meter, the latter are not under control.

2.3 The procedure for dealing with internal escapes is therefore different from that for external escapes.

3. All Gas Escapes — General Procedure

3.1 When advised by a customer of a gas escape, the customer should be told to turn off the gas supply at the customer control valve immediately.

If you are unable to attend to the gas escape within one hour, the customer should be told and advised to contact another CORGI Installer who operates an emergency gas escape service (such as British Gas).

3.2 Test the installation for soundness in accordance with the procedure laid down in Testing for Soundness.

3.3 If the installation is found to be leaking, take the necessary steps to repair leakage. Should the situation arise where the customer control valve cannot readily be turned off, then the priorty of action on site is in this sequence: —

a Open all doors and windows and ventilate area.

b Extinguish all naked lights such as gas, oil and solid fuel fires.

c Do not allow light and power point switches to be turned on or off.

d Evacuate the premises if considered necessary.

e Take steps necessary to close customer control valve and proceed with repair.

3.4 If the installation is found sound, then the smell of gas may be attributable to an external leakage. The priorty of action on site is in this sequence: —

a   Open all windows and doors.

b   Extinguish all naked lights, such as gas, oil and solid fuel fires.

c   Do not allow light switches or power point switches to be turned on or off.

d   Evacuate premises if considered necessary.

e   Check premises opposite and adjacent.

f   Report gas escape immediately to British Gas and remain on site until British Gas personnel arrive.

4. Leakage Location

4.1 In no circumstances, must a naked light be used for illumination or for pinpointing the source of leakage. A soap and water solution must be used for this purpose.

5. Further uses of the "U" Gauge

5.1 Testing for excessive pressure loss due to: —

a   **Partial stoppage.**

The position of the obstruction in the sketch below can be found by taking the working pressure readings along the supply. A loss of 9 mbar from 'A' to 'E' indicates that the obstruction is between 'B' and 'E' since the pressure loss between 'A' and 'C' comparatively small at 3 mbar.

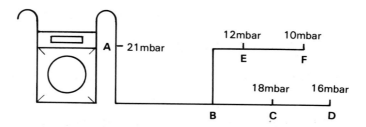

b Inadequate pipe size

The pressure loss due to inadequate pipe size would be regular but excessive. This trouble would occur when an appliance with a high gas rate is added to an existing installation without sufficient thought being given to its effect on existing pipework.

c Overloaded Meter

A gas meter can be checked for overloading by taking the difference between the inlet and outlet pressures when appliances are in operation. This should not exceed 1.25 mbar (5/10 in W.G.).

Adjusting Appliance Pressures

Gas Appliances are designed to their best performance and greater efficiency on a particular gas rate at a definite governor pressure and a fixed injector orifice.

Setting and Checking Constant Pressure (C.P.) Governors

With the gauge attached to a pressure point close to the outlet side of appliance governor, the outlet pressure is altered by varying the tension of the spring. Having set the burner pressure according to the maker's instructions, the gas rate is then checked by timing the test dial hand of the meter with a watch. Some appliances are fitted with pre-set C.P. governors which cannot be adjusted, or no C.P. governors at all, (e.g. domestic cookers).

Note: The Gas Safety (Installation & Use) Regulations 1994 require meter governors to be sealed. Contact your local Gas Region if adjustments are required.

GAS SUPPLY

PROPERTIES OF NATURAL GAS

Occurring naturally in underground accumulations National Gas consists mainly of methane. The analysis of any specific natural gas varies with the particular source from which it is obtained. Methane is always the major constituent but this is mixed with different proportions of other hydrocarbons and inert gases.

Typical natural gas analysis is:

Constituent	Formulae	Percentage by volume
Nitrogen	N_2	2.7
Carbon Dioxide	CO_2	0.6
Methane	CH_4	90.0
Ethane	C_2H_6	5.3
Propane	C_3H_8	1.0
Butane	C_4H_{10}	0.4
		100.0

Odour

Natural gas does not have its own smell but the authorised supplier is required to add a smell to it before distributing the gas to the customer. An odorant is used which usually contains diethyle sulphide and ethyl butyl mercaptan.

Toxicity

Natural gas does not contain carbon monoxide (CO), which by replacing oxygen in the body's blood stream prevents the blood from maintaining life and poisons the organs of the body, so is NON TOXIC. However natural gas contains carbon (0.6 carbon dioxide and **all fuels which contain carbon can produce carbon monoxide in their flue gases if the carbon is not completely burned.** It is therefore essential that we comply with the Gas Safety Regulations which require that any person installing or servicing a gas appliance shall ensure that the:

— means of removal of the products of combustion,

— availability of sufficient air for combustion,

— means of ventilation of the room or space in which the appliance is fitted,

are such that the appliance can be used without danger to any person or property.

Calorific Value

British Standards Institution: BS 1179: Glossary of Terms used in the gas industry defines: Calorific value is the amount of heat liberated by the complete combustion, under specified conditions, of Unit Volume of gas.

All gasses which burn give off heat, the calorific value indicates the heating power. It is a number of heat units which can be obtained from a measured volume of gas. For gases the calorific value is usually expressed:

— Imperial units: CV is the number of British thermal units in each cubic foot of gas (Btu/ft³).

The calorific value of natural gas is approximately 1035 Btu/ft³.

— SI units: CV Megajoules per cubic metre used (MJ/m³ (st)).

The calorific value of natural gas is approximately 39.3 (MJ/m^3 (st)).

NB   North Sea gas has varying characteristics according to the place of origin. After treatment most is supplied dry with a calorific value as shown, this declared CV is changed marginally in some areas.

To convert from Imperial units to SI units, use the following conversion factor:

Btu/ft³ × 0.038 = MJ/m³(st)

To convert from SI units to Imperial units the following conversion factor is used:

MJ/m³ × 26.34 = Btu/ft³

Conversion factors see page 41.

When fuel containing hydrogen is burned water is invariably produced. If this water is condensed, it gives up its latent heat as steam, together with the heat liberated on cooling from its condensation point to the temperature of the calorimeter. The total calorific value (GROSS VALUE) includes this heat. In many cases however this heat is not available for work and for such computations it is eliminated. The value after this deduction is termed the **Net Calorific Value**.

Gas is sold on the basis of its gross calorific value and because customers pay for the gas they use measured in heat units, the authorised gas supplier has to declare the calorific value of its supply, this is monitored at official testing stations by gas examiners appointed by the ministry.

Gross Calorific Value

The gross calorific value of a fuel gas at constant pressure is the number of units of heat liberated by the complete combustion in air of unit volume of the gas saturated with water vapour, both air and gas being at 15 degrees centigrade (60 degrees

Fahrenheit) while the effluent gases are cooled to 15°C and all the water contained in them in excess of that carried as vapour by the entrant air and gas is condensed to liquid state and cooled at 15°C. By complete combustion it is meant that there shall be no products other than carbon dioxide, water, sulphur dioxide and nitrogen.

Net Calorific Value

The net calorific value of a fuel gas determined at constant pressure is the number of heat units liberated by the complete combustion in air of a unit volume of gas saturated with water vapour, both air and gas being at 15°C (60°F) while the effluent gases and vapour are cooled to 15°C and all water contained in them is assumed to remain in the vapour state.

Simply the total amount of heat obtainable from the gas is, in fact the Gross Calorific value, if however the water vapour in the products of combustion is not allowed to consdense into water, the amount of heat obtained is the Net Calorific Value.

Relative Density

BS 1179: Relative Density is the Ratio of the mass of unit volume of dry gas to that of unit volume of dry air under the same conditions of temperature and pressure.

Every substance has weight or mass, including gas. It is necessary, for various reasons, to compare weight of gases and to do this a comparison is made of densities. In order to compare densities they are related to a standard substance; for gases the standard is air.

This relationship between density of a substance and the density of a standard is known as relative density (or specific gravity).

The Relative Density (specific gravity) **of Natural Gas is in the region of** 0.58 so it is about half the weight of the same volume of air.

Wobbe Number

BS 1179: The heat release when a gas is burned at a constant gas supply pressure expressed by:

$$\frac{\text{gross calorific value of the gas}}{\sqrt{\text{relative density of the gas}}}$$

The Wobbe number presents a measure of the heat release when a gas is burned at a constant gas supply pressure. The heat release is then directly proportional to the orifice area and the Wobbe number.

The amount of heat which a burner will give is dependent on:

1.   The amount of heat in the gas as given by its calorific value.

2.   The rate at which the gas is being burned which is dependent on:

   2.1   The size of the injector.

   2.2   The pressure in the gas.

   2.3   The relative weight of the gas, indicated by the relative density.

Factors dependent on gas   —   Calorific value (1).
                                 Relative density (2.3).

Factors depending on the
appliance   —   Size of injector (2.1).
                            —   Pressure of the gas (2.2).

Since the factors in group 2 are fixed by the design of the appliance, the only alteration in heat output would be brought about by changes in the group 1 factors. That is, by changes in the characteristics of the gas, the calorific value, and the relative density. The Wobbe number links these two characteristics and is obtained by dividing the CV by the square root of the relative density. It is essential to ensure that the heat outputs of appliances are kept reasonably constant, therefore the Wobbe number of the gas must be maintained within close limits.

Natural Gas with Calorific Value of 1035 BTU/ft^3 and a Relative Density of 0.58 would have a Wobbe Number of 1357. In MJ/m^3 this would be 51.56 (1 MJ/m^3 = 26.34 BTU/ft^3).

Gas Classification

The various fuel gases used differ greatly in properties, with the result that an appliance designed and adjusted to burn one type of gas would be totally unsatisfactory if supplied with different gas. The most important property of gas is the Wobbe number and it has been agreed by the International Gas Union that fuel gases be classified according to their Wobbe number into three families.

Gas Families

BS 1179: A range of gases characterised by having a Wobbe number within specified limits.

First Family   : Gas of Wobbe number 24.4 MJ/m³ to 28.8 MJ/m³
                    under standard conditions.

Second Family : Gas of Wobbe number 48.2 MJ/m³ to 53.2 MJ/m³
under standard conditions.

Third Family : Gas of Wobbe number 72.6 MJ/m³ to 87.8 MJ/m³
under standard conditions.

Table 7.

| Family | Approx. Wobbe No. Range. | | Type of Gas |
	MJ/m³	Btu/ft³	
1.	24.4 — 28.8	640-750	Manufactured
2.	48.2 — 53.2	1250 — 1380	Natural
3.	72.6 — 87.8	1890 — 2300	LPG

Appliances are designed to operate on gas of a particular family.

Gas Groups

BS 1179: a subdivision of a family of gases.

Even within a family the range of gas properties is too wide to be accommodated by a single appliance setting. The variation in Wobbe number is such as to cause an unacceptable variation in appliance heat input. Each family is therefore subdivided into groups.

i.e. First Family

(a) town gas   (b) coke oven gas   (c) LPG/air

Air Requirements

Fuel gases burn when they are ignited and combined with oxygen. The atmosphere consists of about 21% oxygen and 79% nitrogen which means that if a gas flame is allowed to burn freely in the open it obtains the oxygen it needs from the surrounding air. For each cubic metre (m³) of gas burned the amount of air required for Natural gas: 9.7m³.

Gas Modulus

BS 1179: The Ratio $\dfrac{\sqrt{\text{injector pressure}}}{\text{Wobbe number}}$

The gas modulus is a numerical expression which relates to the heat output from a burner with pressure required to provide a satisfactory amount of aeration. It gives a figure which indicates how aeration and heat loading conditions may be maintained when changing from one gas to another. The modulus is obtained by dividing the square root of pressure by the Wobbe number.

Using the modulus shows that to change from a manufactured gas with a Wobbe index of 27.2 supplied at a pressure of 6.23 m bar to a Natural gas with a Wobbe index of 49.6 required the pressure to be increased to 20.62 m bar to maintain the same operating conditions.

Flame Speed

BS 1179: The rate of linear propagation of flame through a gas mixture.

The flame speed of gas is measured in feet or metres per second. The flame speed of Natural gas is:

1.1 ft/s or 0.36 m/s.

Ignition Temperature

Gases need to be ignited before they will burn. This is done by heating the gas until it reaches a sufficiently high temperature to burst into flame or ignite. The ignition temperature of Natural gas is:

1300 degrees Fahrenheit or 704 degrees celsius.

CONVERSION FACTORS

Cubic ft of gas to therms:

$$\text{No. of therms} = \frac{\text{No. of ft}^3 \times \text{calorific value}}{100\ 000}$$

Example: The number of therms in 5000 ft³ of gas (c.v. 1000)

$$= \frac{5000 \times 1000}{100\ 000} = 50$$

Imperial to Metric Units	
1 inch	= 25.4 mm = 2.54 cm
1 foot	= 304.8 mm = 0.3048 m
1 yard	= 914.4 mm = 0.9144 m
1 mile	= 1609 m = 1.609 km
1 square inch	= 645.2 mm² = 6.452 cm²
1 square foot	= 929.0 cm²
1 square yard	= 0.8361 m²
1 acre	= 4047 m² = 0.4047 ha
1 cubic inch	= 16 390 mm³ = 0.016 39 dm³
1 cubic foot	= 0.028 32 m³ = 28.3 dm³
1 cubic yard	= 0.7647 m³
1 pint	= 0.5683 litre = 568.3 ml
1 quart	= 1.137 litre
1 gallon	= 4.546 litre = 4.546 kg of water
1000 gallons	= 4.546 m³
1 grain	= 0.064 80 g
1 ounce	= 28.35 g
1 pound	= 453.6 g = 0.4536 kg
1 ton	= 1016 kg = 1.016 tonne
0.1 inches water gauge	= 0.25 mbar
1.0 inch water gauge	= 2.5 mbar
1 atmosphere	= 1.013 25 bar
British thermal unit	= 1055 J
1000 Btu	= 1.055 MJ
1 therm	= 105.5 MJ
1000 Btu/h	= 0.2931 kW = 1.055 MJ/h
1 horsepower	= 745.7 W = 2.685 MJ/h
1 Btu/sft³	= 0.037 96 MJ/m³ (st)*
1 Btu/ft³ (dry)	= 0.037 23 MJ/m³ (st)*
1 Btu/lb	= 2326 J/kg

Metric to Imperial Units

1 millimetre	= 0.039 37 in
1 metre	= 39.37 in = 3.281 ft = 1.094 yd
1 kilometre	= 0.6214 mile = 1094 yd
1 square millimetre	= 0.001 550 in²
1 square metre	= 1.196 yd²
1 are	= 119.6 yd² = 0.0247 acre
1 hectare	= 11 960 yd² = 2.471 acre
1 cubic millimetre	= 0.000 061 in³
1 cubic decimetre	= 61.02 in³ = 0.035 31 ft³
1 cubic metre	= 35.31 ft³ = 1.308 yd³ = 220.0 gal
1 millilitre	= 0.001 760 pint
1 litre	= 1.760 pint = 0.2200 gal = 0.035 ft³
1 gram	= 0.035 27 oz = 15.43 grains
1 kilogram	= 2.205 lb
1 tonne	= 2205 lb = 0.9842 ton
1 millibar	= 0.4 in wg
1 bar	= 14.50 lbf/in²
1 joule	= 0.000 947 8 Btu
1 megajoule	= 947.8 Btu
100 megajoules	= 0.9478 therm
1 kilowatt	= 3.6 MJ/h = 3412 Btu/h
	= 1.341 horsepower
1 kilowatt hour	= 1 kWh = 3.6 MJ
1 megajoule per hour	= 0.2778 kW = 947.8 Btu/h
1 MJ/m³	= 26.34 Btu/ft³

*See Calorific Values, p. 20.

General

1 gallon	= 0.16 ft³ = 10 lb of water
1 ft³ water at 62°F and 30 in Hg	= 62.3 lb = 6.25 gal = 28.3 litres
1 horsepower (hp)	= 33 000 ft lb/min = 42.41 Btu/min
	= 2544 Btu/h
Normal air pressure	= 760 mm Hg = 29.92 in Hg
	= 1013.25 mbar
Normal Temperature and Pressure (NTP)	= 0°C, 760 mm Hg (mercury)
Imperial Standard Conditions (STC)	= 60°F, 30 in Hg (i.e. 1013.7405 mbar) saturated
Metric Standard Conditions (MSC) (st)	= 15°C, 1013.25 mbar dry
1 pound per square inch	= 27.7 in water = 2 in Hg
1 inch of mercury	= 13.57 in water = 0.49 lb/in²
Specific heat of water	= 4.187 kJ/kg°C

Converting Temperature Readings

The relationship may be expressed.

$$\frac{(°F - 32)}{9} = \frac{°C}{5}$$

or $°C = \frac{5}{9} (°F - 32)$ or $°F = \frac{9 \times °C}{5} + 32$

Table 8.

°C	°F	°C	°F	°C	°F
-273.1	-459.6	25	77	85	185
-161.7	-259	30	86	90	194
- 40	- 40	35	95	95	203
- 20	- 4	40	104	100	212
- 17.8	0	45	113	110	230
- 10	14	50	122	150	302
- 5	23	55	131	200	392
0	32	60	140	300	572
5	41	65	149	400	752
10	50	70	158	500	932
15	59	75	167	1000	1832
20	68	80	176	1500	2732

Note — The metric Celsius temperature scale is the centigrade scale renamed. °C remains the accepted abbreviation for temperature value. Celsius is likely to remain in common usage even after the official adoption of the SI Kelvin scale.

Modifying Gas Appliances

Most appliances which were designed to burn family 1 gases can be converted or modified to work on family 2 gases and vice versa.

As an example, two typical gases are:

1. Family 1   Calorific value (CV)   = 18.9 MJ/m³
   Relative density (SG)   = 0.475
   (specific gravity)
   Wobbe number $\dfrac{18.9}{\sqrt{0.475}}$ = 27.4

   Air required to burn
             1 m³   = 4.3 m³

   Volume of products from
             1 m³   = 5.0 m³

2. Family 2   Calorific value   = 38.5 MJ/m³
             Relative density   = 0.601

$$\text{Wobbe number} \frac{38.5}{\sqrt{0.601}} = 49.7 \text{ m}^3$$

Air required to burn
$$1 \text{ m}^3 = 9.8 \text{ m}^3$$

Volume of products from
$$1 \text{ m}^3 = 10.8 \text{ m}^3$$

Comparison

CV: The calorific value of family 2 is about twice that of family 1 so only half the volume of family 2 gas is required to give the same heat input.

SG: The relative density (specific gravity) of family 2 is slightly higher than that of family 1 so the family 2 Wobbe number is not quite twice that of family 1. This affects the gas modulus figure and means that a much higher pressure will be required at the injector to maintain satisfactory aeration.

Air Requirement

Family 2 needs about twice as much air as family 1 but since half the volume of family 2 is used then the air inlets should be adequate.

Products of Combustion

Family 2 produces about 20% greater volume of burnt gases than twice the amount produced by family 1. So if half the volume of family 2 is used it will result in about 10% greater volume of products than the original family 1. If the flue way of the appliance were designed to allow at least this amount of excess air, combustion would still be satisfactory. If the allowance for excess air were less, then the appliance must have its gas rate reduced to ensure satisfactory combustion. This is called down rating the appliance.

The actual pressure and new injector diameter required can be calculated using the gas modulus formulae.

i.e. if the appliance has an injector of 3.5 mm working on a pressure of 5 m bar on family 1 gas, the new pressure and diameter are as follows:

$$\text{Gas Modulus} \frac{\sqrt{\text{Pressure}}}{\text{Wobbe Number}}$$

or $\sqrt{\dfrac{P}{W}}$

or $\sqrt{\dfrac{P1}{W1}} = \sqrt{\dfrac{P2}{W}}$ to maintain same conditions

or $\sqrt{\dfrac{5}{27.4}} = \sqrt{\dfrac{P2}{49.7}}$

$\sqrt{P2} = \sqrt{\dfrac{5 \times 49.7}{27.4}}$

$= 4.06$

Therefore P2 = 16.5 m bar.

To find the injector size use the formulae: $Q = 0.467 \times A \times C\sqrt{\dfrac{P}{S}}$

usually written $Q = 0.036\ d^2\ C\sqrt{\dfrac{P}{S}}$

Where Q = volume of gas in cubic metres per hour (m³/h).
       A = area of orifice in square millimetre (mm³).
       D = diameter of orifice in millimeters (mm).
       P = pressure behind the injector in millibars (mbars).
       S = specific gravity of gas.
       C = the coefficient of discharge.

Assume the coefficient of discharge is the same for both injectors, then since the gravity of gas, Q must remain the same.

$Q = 0.036\ Cd_1{}^2\sqrt{\dfrac{P1}{S1}}$

$= 0.036\ Cd_2{}^2\sqrt{\dfrac{P2}{S2}}$

divide both sides by 0.036 to obtain:

$$3.5_2 \sqrt{\frac{5}{0.475}} \;=\; d_2^2 \sqrt{\frac{16.5}{0.601}}$$

which gives
$$d_2^2 \;=\; 7.57$$

$$d2 \;=\; 2.75 \text{ mm}$$

So with family 2 gas, an injector with a diameter of 2.75 mm and a working pressure of 16.5 mbar is required.

Comparison of Properties of Typical Gases

Property	Units	Natural Gas	Manufactured Gas	Commercial Propane
Calorific value	Btu/ft³	1034	500	2590
	MJ/m³	38.5	18.63	97.3
Specific gravity	air = 1	0.58	0.47	1.5
Wobbe number	Btu/ft³	1330	730	2140
	MJ/m³	49.6	27.2	798
Air required	volume/volume	9.75	4.3	23.8
Flammability limits	% gas in air	5 to 15	4 to 40	1.6 to 7.75
Flame speed	ft/s	1.12	3.28	1.4
	m/s	0.36	1.0	0.46
Ignition temperature	°F	1300	1100	986
	°C	704	593	530

Table 9.

46

PRESSURE AND FLOW

Gas Cocks

The simplest and most common type of control is the gas cock, which is used to shut off the supply of gas to the meter or to individual gas appliances. The cock consists of a tapered plug having a slot through it, which fits into a tapered body. The two connecting surfaces of the plug and tapered body are machined to give a gas-tight seal and these surfaces should be smeared with grease to act as a lubricant. The cock is opened or closed by turning the plug through an angle of 90°. The drawing below shows a section through a main control cock which incorporates a union to facilitate the removal of the meter (Fig. 17).

For safety reasons, a drop-fan cock may be used as shown. The fan is hinged to the plug and has lugs which hold the fan upright when the gas is on. When the cock is turned off, the fan falls and the lugs engage with a slot in the tapered body of the cock. The plug in this off position cannot be turned on until the fan is deliberately held upright and this prevents the cock from being turned on accidentally.

An even safer type of control is known as the plug-in safety cock. This type of cock cannot be turned on without first inserting the outlet pipe to the body of the cock. When the outlet pipe plug is inserted and turned to engage the lugs, two pins on the plug also engage a groove, and a gas-tight seal is made between the two components by compressing the main spring, which forces the two conical surfaces together (Fig. 18).

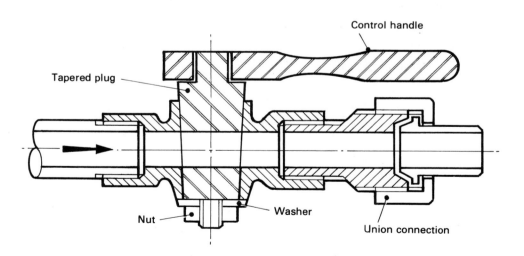

Figure 17. Main Control Cock

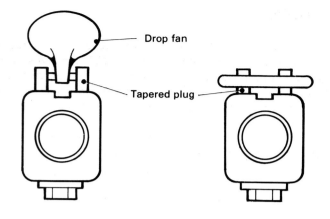

Drop Fan Safety Cock

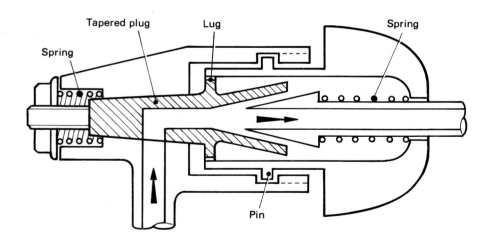

Plug-in Safety Cock

Figure 18. Gas Safety Cocks

GAS GOVERNORS

Natural gas rises to the North Sea well-heads at pressures in the region of 100 bar (1500 lb/in²), and arrives at the customer's meter at 21 mbar (8.4 in wg). It is subjected to numerous fluctuations due to temperature changes, friction, altitude, atmospheric pressure, etc, but is of paramount importance that the gas pressure remains constant at the appliance. The governor carries out this task.

Governors come in all shapes, sizes, designs and complexities, one thing that does not vary, is their function — the automatic limitation of the gas pressure or flow rate.

A service governor is fitted on a gas service pipe, that is, upstream of the meter control.

A meter governor is fitted between the meter control and the meter, that is, downstream of the meter control. So, as defined by the Gas Safety Regulations, a service governor is on the service and meter governor is part of the meter installation.

Internal installation pipework starts after the meter installation.

Most domestic meters are supplied through low pressure services and are fitted with meter governors. The governor is generally connected directly to the inlet of the meter. To rid the gas flow of impurities it has been found necessary to incorporate a filter in the governor.

As a general rule the specification of meter governors falls within the following categories:

Inlet and outlet connections	25 mm (1 in)
Outlet pressure range	12.5 to 25 mbar (5 to 10 in wg)
Maximum inlet pressure	75 mbar (30 in wg)
Maximum gas rate	14m³/h (500 ft³/h)

Need for a Meter Governor

Pressures vary with changes in demand for gas. When very few appliances are in use pressures are at their highest. As more and more gas is consumed, the loss of pressure is greater. If the customer's appliances are to work efficiently there are limits below which the pressure must not be allowed to fall. On the safety side, British Gas is required by statute to maintain a safe minimum pressure of 12.5 mbar (5 in wg) in the mains at all times.

To maintain adequate supplies the pressure must be controlled at all times, too high pressure is equally as bad as one which is too low, as excessive gas rates on an appliance can affect the combustion and may present hazards. To ensure that each appliance is supplied with gas at the pressure for which it was designed, it is necessary to control the pressures within fairly close limits, and, once set, they must be kept constant. The meter governor is usually the last in a series of valves, or variable restrictors, designed to achieve this condition. The 'constant' requirement of all these various control valves is that they should adjust automatically so that they will always provide an adequate supply of gas at a constant outlet pressure, irrespective of changes in the inlet pressure or in the rate of gas flow.

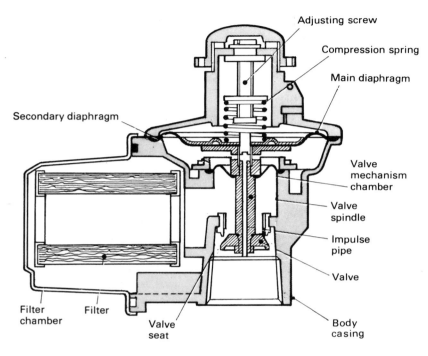

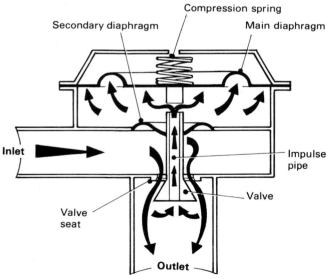

Figure 19. Gas Governor Mechanism

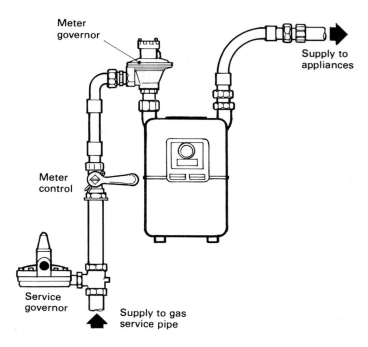

Figure 20. The position of the Service Governor and Meter Governor in relation to the Meter Control.

Development and Construction of Meter Governors

Modern day governors use 'spring loading' which operates on the principle that a compressed spring possesses potential energy and is therefore capable of exerting a force. The major advantage of spring loading is that it enables the governor to be fitted with the diaphragm in any position, even completely vertical.

Operation of a Meter Governor

The gas at inlet pressure enters the governor at the inlet port and is channelled, through the gasways in the body casting, to the outer periphery of the filter chamber. The gas then passes through the filter into its core, and from the core into the valve mechanism chamber.

With no gas below the main diaphragm in the diaphragm chamber, the compression spring is extended – holding the diaphragm down and, as the valve mechanism is an assembly, the valve off its seat. As the gas passes between the valve and seat towards the outlet it undergoes a restriction. Some of this gas, now at a reduced pressure, is fed via the impulse pipe (a passage within the valve spindle) to the diaphragm chamber below the main diaphragm.

As the chamber fills with gas the resultant pressure overcomes the spring resistance and raises the diaphragm, and thus lifts the valve assembly, closing the gap between the valve and its seat and further reducing the outlet gas pressure. The 'constant' outlet pressure is adjusted by means of the adjusting screw acting on the spring to hold the valve assembly and main diaphragm in equilibrium.

The secondary diaphragm is to compensate for the effects of varying inlet pressures that may act above the valve as the gas passes from the valve mechanism chamber to the outlet port.

Under no-gas-flow conditions the valve locks-up. This is the term applied to the condition which occurs when the gas is turned off on the outlet side of the governor causing the pressure below the diaphragm to increase slightly so closing the valve completely.

Operation of a Simple Governor

In principle all governors operate in a similar way. Fig. 21 shows a sectional elevation through a simple constant pressure governor, and details the main parts.

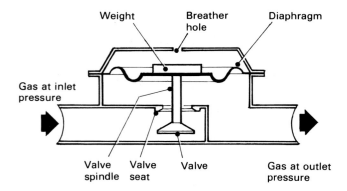

Figure 21. A simple Constant Pressure Governor

The forces which are acting on the valve and diaphragm are shown below. The main forces on the diaphragm are:

Downward force = Weight or the forces exerted by the mass of the weight (or spring pressure).

= W

Upward force = Outlet pressure x the effective area of the diaphragm.

= OP x A

When these two forces balance, the valve is held off its seating in a state of equilibrium. The amount by which the valve is open will be that which will permit sufficient gas to pass through to maintain the outlet pressure at a value which will just maintain the balance. So:

$$OP \ x \ A \ = \ W \qquad \text{or} \qquad OP \ = \ \frac{W}{A}$$

IP = Inlet Pressure
OP = Outlet Pressure
 W = Weight
 A = Effective area of Diaphragm

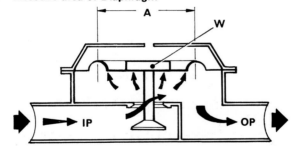

Figure 22. Forces acting on the Valve Diaphragm

The diagrams (Fig. 23a, b, c) show how changes in operating conditions can affect the amount of valve opening.

(a) the valve is only partly open, supplying a small amount of gas at an outlet pressure of OP.

If the demand for gas is greater, gas will be taken from under the diaphragm more quickly than it is being supplied through the valve seating. OP will be reduced, the balance upset, and the downward force W will be greater than the upward force OP x A. So the weight (or spring) will push the valve down until the opening is large enough to supply sufficient gas to the appliance(s) and again bring the outlet pressure back up to:

$$OP = \frac{W}{A}$$

The valve will now be in the position shown at (b). Further demands could result in the valve opening to the position shown in (c). The position of the valve is dependent upon the flow of gas required, it also depends on the inlet pressure (IP). With the valve in the position shown in (b), any increase in inlet pressure would tend to push the gas quickly past the valve and the outlet pressure would increase also. Once again the balance would be upset, but the stronger force would now be the upward one (OP x A), and the valve would close slightly as in (a), so offering greater resistance to the higher pressure. The reverse of this would happen if the inlet pressure were to fall. The valve would open, as in (c), to maintain a constant outlet pressure. The governor can only operate satisfactorily when the inlet pressure is above that required at the outlet.

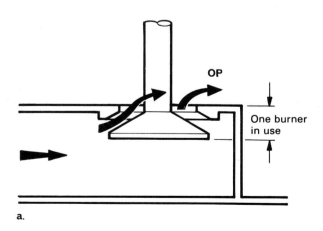

Figure 23. Valve partly open

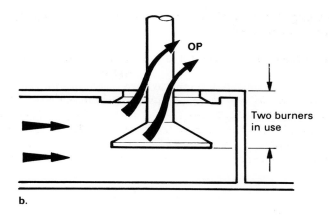

b.

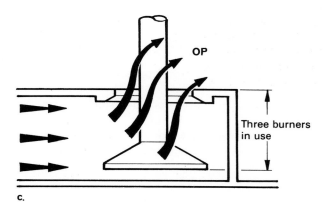

c.

Figure 23. The effect of increased gas rate and subsequent falling outlet pressure on the valve position

COMBUSTION

Combustion Process

Every gas appliance is a device designed to burn gas to produce heat. Natural gas is mostly methane, and during combustion reacts with oxygen in the air. It is a chemical reaction and may be written in the form of equations.

i.e. Combustion of methane.

$$CH_4 + 2O_2 = CO_2 + 2H_2O$$

Almost incidental to this chemical reaction is the liberation of heat energy, but it must be realised that nothing is actually being consumed — it is the transformation of the chemical constituents which produce the heat.

Methane, CH_4, burns in combination with oxygen, O_2, in the air. In doing so it produces carbon dioxide, CO_2, and water vapour, H_2O.

With all equations both sides balance.

— There is one carbon atom either side.

— There are four hydrogen atoms on each side.

— There are four oxygen atoms on each side.

Since there is a direct relationship between the number of molecules and the volume of gases, the equation shows the volumes of gases included.

CH_4	+	$2O_2$	=	CO_2	+	$2H_2O$
1 volume methane	requires	2 volumes oxygen	and gives off	1 volume carbon dioxide	and	2 volumes water vapour

This is true for any volume of methane.

Natural gas is a mixture of gases and their constituents are given below.

The chemical equation for their combustion are as follows:

Nitrogen : This is not flammable and takes no part in combustion, hence it is called inert.

Carbon Dioxide : This is not flammable. It is actually a product of combustion.

Methane : $CH_4 + 2O_2 = CO_2 + 2H_2O$

Ethane	: $2C_2H_6 + 7O_2 = 4CO_2 + 6H_2O$
Propane	: $C_3H_8 + 5O_2 = 3CO_2 + 4H_2O$
Butane	: $2C_4H_{10} + 13O_2 = 8CO_2 + 10H_2O$

Air Requirements

It is possible to find out the combustion air requirements for a gas by first calculating the amount of oxygen required, as indicated by the combustion equation. In natural gas there is 90m³ of methane in each 100m³ of the gas. The equation for the combustion of methane is:

	CH_4	+	$2O_2$	=	CO_2	+	$2H_2O$
or	1 volume	+	2 volumes	=	1 volume	+	2 volumes

So each 1m³ of methane requires 2m³ of oxygen and produces 1m³ of carbon dioxide and 2m³ of water vapour.

Therefore 90m³ of methane will require 180m³ of oxygen and produce 90m³ of carbon dioxide and 180m³ of water vapour.

The oxygen requirements for natural gas therefore are as shown.

Table 10.

Constituent	% by volume	Chemical Equation for Combustion	Volumes of Oxygen required to burn
N_2	2.7	—	
CO_2	0.6	—	
CH_4 (methane)	90.0	$CH_4 + 2O_2 = CO_2 + 2H_2O$	$\frac{90}{1} \times 2 = 180.0$
C_2H_6 (ethane)	5.3	$2C_2H_6 + 7O_2 = 4CO_2 + 6H_2O$	$\frac{5.3}{2} \times 7 = 18.6$
C_3H_8 (propane)	1.0	$C_3H_8 + 5O_2 = 3CO_2 + 4H_2O$	$\frac{1}{1} \times 5 = 5.0$
C_4H_{10} (butane)	0.4	$2C_4H_{10} + 13O_2 = 8CO_2 + 10H_2O$	$\frac{0.4}{2} \times 13 = 2.6$
TOTAL	100		206.2

So 1m³ of natural gas requires 2.062 m³ of oxygen — 2m³.

Since the atmosphere consists of 21 per cent oxygen the air requirements for the **complete combustion** of 1 m³ of natural gas is:

$2.062 \times \dfrac{100}{21} = 9.8$ m³ of air, approximately 10m³.

Products of Combustion

The products of combustion will contain the carbon dioxide and water vapour, resulting from the chemical reaction plus the inert nitrogen, however it doesn't follow that because 1 m³ of natural gas requires 9.8 m³ of air, the volume of products will be 1 + 9.8 = 10.8 m³ of products of combustion.

For example in the combustion of propane:

$$C_3H_8 + 5O_2 = 3CO_2 + 4H_2O$$
$$1 \text{ volume} + 5 \text{ volumes} = 3 \text{ volumes} + 4 \text{ volumes}$$

The volumes are not equal whether we use O_2 or air volumes, but what the equation does show is that 1 m³ of propane combined with 5 m³ of oxygen will produce 3 m³ of carbon dioxide and 4 m³ of water vapour. The amount of the products of combustion produced in each case can be calculated from the combustion equations.

Products of combustion — Natural Gas

Constituent	% by volume	Chemical Equation for Combustion	Products of Combustion	
			Carbon Dioxide	Water Vapour
N_2	2.7	—	—	—
CO_2	0.6	—	06	—
CH_4	90.0	$CH_4 + 2O_2 = 2CO_2 + 2H_2O$	$\frac{90}{1} \times 1 = 90.0$	$\frac{90}{1} \times 2 = 180.0$
C_2H_6	5.3	$2C_2H_6 + 7O_2 = 4CO_2 + 6H_2O$	$\frac{5.3}{2} \times 4 = 10.6$	$\frac{5.3}{2} \times 6 = 15.9$
C_3H_8	1.0	$C_3H_8 + 5O_2 = 3CO_2 + 4H_2O$	$\frac{1}{1} \times 3 = 3.0$	$\frac{1}{1} \times 4 = 4.0$
C_4H_{10}	0.4	$2C_4H_{10} + 13O_2 = 8CO_2 + 10H_2O$	$\frac{0.4}{2} \times 8 = 1.6$	$\frac{0.4}{2} \times 10 = 2.0$
TOTAL	100		105.8	201.9

Table 11.

To provide the 2.06 m³ of oxygen required to burn 1 m³ of natural gas we need 9.8 m³ of air. The difference, 9.8 - 2.06 = 7.74 m³ is nitrogen. So when 1 m³ of natural gas is burned, the total volume of gases which are leaving the combustion chamber of the appliance is:

Carbon dioxide	1.058
Water vapour	2.019
Nitrogen from gas	0.027
Nitrogen from air	7.74
TOTAL	10.844 m³ approximately 11 m³

58

Incomplete Combustion

In the reaction zone of a flame constituent gases are broken down and recombine to form other compounds before being burned to the final products of combustion, CO_2 and H_2O (carbon dioxide and water vapour).

These intermediate substances include carbon, carbon monoxide, alcohols and aldehydes. If combustion is interupted at this stage it is **incomplete combustion** and these products can be released from the flame.

Of the products of incomplete combustion, carbon monoxide is the most dangerous. **It is a toxic gas.** A concentration of only 0.4 per cent in the air can be fatal within a few minutes, even if death does not occur there may be serious brain damage. Carbon itself is not dangerous. It can be seen in the flame as a yellow zone. If this part of the flame touches a cold surface carbon is deposited in the form of soot.

Soot is a nuisance and its formation in the narrow passages of the combustion chamber can restrict the flow of products of combustion, which in turn reduces the flow of air to the flame which could result in a more serious deposit of soot and carbon monoxide can be formed. Alcohols are quickly oxidised to become aldehydes which have a characteristic smell and often give the indication that incomplete combustion is taking place. Although they are poisonous they are a nuisance rather than a danger causing irritation to the eyes.

Anything that interferes with the combustion process may result in the release of products of incomplete combustion. The two main causes are chilling the flame and starving it of oxygen.

Chilling occurs when a flame touches a cold surface, cold compared to the flame. If the surface cools the gases in the flame to below their ignition temperature, some of them will not be completely burned and may release the products of incomplete combustion.

The effects can be alarming if the inner cone of a pre-aerated flame is broken by the cold surface because this inner cone contains unburned gases.

Chilling can also result if the flame is exposed to cold air or to a draught which may tend to lift the flame off the burner. If the flame is under aerated and yellow-tipped and touches a cold surface, soot will be deposited causing interference.

Lack of oxygen produces carbon monoxide and may be caused by restricting the amount of air to the burner or by a shortage of oxygen in the air.

The air supply to the burner may be restricted either by a blockage of the air inlet to the appliance or by a blockage to the flue outlet. Both have the same effect on the flow of air over the burner.

Dirt, fluff or soot on or in the burner or mixing tube will restrict air to the flame even though the total amount of air required is entering the appliance. A lack of oxygen in the air may mean a shortage of air to the burner. Air has approximately 20% oxygen content with a negligible amount of carbon dioxide. If the gas appliance is located in a room without proper ventilation it uses up the oxygen and gives out carbon dioxide and water vapour thus reducing the proportion of oxygen compared to other gases affecting combustion. This lack of oxygen is called **vitiation**. Air is made impure (vitiated). Adequate ventilation is required to prevent air becoming vitiated.

A flame inside an appliance will become larger in its attempt to reach for oxygen which may cause it to touch the combustion chamber resulting in bad combustion. When flames touch each other, or a cold surface, they are said to **impinge**. Flame impingement will normally cause poor combustion either by chilling or by one flame depriving the other of oxygen.

Combustion Standards

PART E Paragraph 26 (1) ACOP of the Gas Safety Regulations states that no gas appliance shall be installed unless:

— the means of removal of the products of combustion from the appliance,

— the availability of sufficient supply of air for the appliance for proper combustion,

— the means of ventilation to the room or internal space in which the appliance is used,

are such as to ensure that the appliance can be used without constituting a danger to any person or property.

Even with this legal requirement, because of the dangers of poor combustion, minimum standards of combustion have been laid down by the British Standards Institution. All 'approved' appliances must conform to this specification, BS 5258.

Standards in Britain are based on the ratio of the amount of carbon monoxide produced to the amount of carbon dioxide produced. **The ratio $CO:CO_2$ must not exceed 0.02.** Under laid down conditions.

Effect of Carbon Monoxide on Adults

% Saturation of Haemoglobin with Carbon Monoxide	Symptoms
0 to 10%	No symptoms.
10 to 20%	Tightness across the forehead. Yawning.
20 to 30%	Flushed skin, headache, breathlessness and palpitation on exertion. Slight dizziness.
30 to 40%	Severe headache, dizziness, nausea. Weakness of the knees, irritability. Impaired judgement.* Possible collapse.
40 to 50%	Symptoms as above with increased respiration and pulse rates. Collapse on exertion.*
50 to 60%	Loss of consciousness, coma.
60 to 70%	Coma, weakened heart and respiration.
70% and above	Respiratory failure and death.

Table 12.

% Volume of Carbon Monoxide in Air	% Saturation of Haemoglobin with Carbon Monoxide
0.01%	4% in 1½ hours. About 15% maximum with indefinite exposure.
0.03%	10% in 1½ hours. 20% in 4 to 5 hours.
0.05%	20% in 1½ hours. 40% in 4 to 5 hours.
0.4%	60% in a few minutes.

Table 13.

*These two factors explain why CO poisoning is frequently fatal.

1. It impairs mental ability so that a person may be brought to the verge of collapse without realising that anything is wrong.

2. Any sudden exertion would then cause immediate collapse, and an inability to escape from the situation.

FLAMES

Introduction

Flames come in all shapes and sizes and are used in a wide variety of applications, in fact flames now provide the vast proportion of energy we use in everyday life. A flame is produced when a mixture of gas and oxygen is burned, it is a zone where chemical reactions are taking place. Natural gas is formed by geological forces acting on decaying plant and marine life which died many millions of years ago. When we burn this gas we rapidly reverse this process; the chemical bonds are rearranged to give carbon dioxide and water vapour, and the solar and geothermal energy that is locked within the fuel is ours to use as heat. In any flame there are intermediate stages of combustion during which other chemicals are produced. As the gas and air diffuse in the flame, they are heated causing the original constituents of the gas to dissociate forming differet compounds of carbon, hydrogen and oxygen. These may be alcohols or aldehydes and there may be some free carbon and carbon monoxide present. By the time the flame has taken in its full requirement of oxygen all the intermediate substances will have been oxidised to form the final products, carbon dioxide and water. So what is a flame? There is yet no complete definition, but some answers lie in the properties of flames which man has long known: fire spreads and heat is released — flames are self sustaining exothermic processes.

There are many different sizes and shapes of flames, each suitable for a particular purpose. Flames can be divided into two broad categories according to the type of burner on which they are formed.

    — Post aerated flames.

    — Pre-aerated flames.

A post aerated flame, sometimes termed non-aerated flame, gets all its air after it has left the burner and a pre-aerated flame gets some (or all) of its air before it leaves the burner.

Post-aerated flames

Also known as 'neat flame'; 'luminous flame' or 'non aerated flame' in which neat gas issues from the burner ports and all the air used in combustion is supplied from surrounding atmosphere. Used extensively in the manufactured town gas era but found not to be satisfactory on natural gas and their substitutes.

At low gas pressure the flame is ragged and shapeless with a large luminous zone.

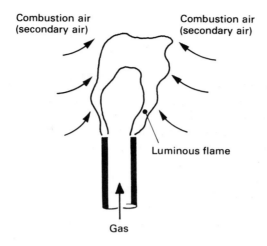

Figure 24. A post-aerated flame

Increasing the pressure can make the air mix with the gas quicker and give a neatly shaped stable flame, this can only be done for gases with a high flame speed like manufactured gas.

With a gas of low flame speed, like natural gas, the flame is blown off the jet and disappears when the pressure is increased, hence their decline, however post-aerated burners have been designed specially for natural gas incorporating some means of keeping the flame alight on the jet, usually by means of a small retention flame supplied with gas at a lower pressure than the main flame.

Pre-aerated flames (aerated flame)

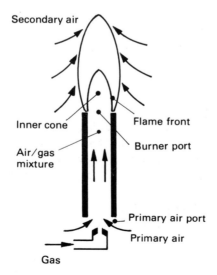

Figure 25. A pre-aerated flame

The flame most commonly used in virtually all of the 40 million domestic gas appliances is the one produced by a burner in which some of the air required for combustion is mixed with the gas before it is burned, known as partially aerated or bunsen burner type, about 50 per cent of the air required for complete combustion is pre mixed with the gas.

The air which is added before combustion is called **primary aeration** and the air needed to complete the combustion is obtained from around the flame itself and is called **secondary air**.

The characteristic of a pre-aerated flame is a well defined inner green/blue cone inside an outer blue/purple flame. (The shape of this flame can be regulated to a desired shape and size).

To understand why the flame takes on this shape it is necessary to look at a simple burner (Fig. 26).

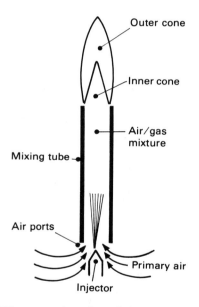

Figure 26. Diagram showing the parts and operation of an aerated (Bunsen) burner

Gas is forced, by pressure, out of a jet placed centrally at the end of a tube. The gas injected into the tube draws in the primary air and pushes it up the tube, mixing it with the gas on the way. The mixture is lit at the top end of the tube. All the holes in burners are called 'ports'. So air is drawn in at the 'primary air ports' and the mixture burns at the 'burner port'. The tube of the burner is called the 'mixing tube' and the jet supplying the gas is an 'injector'. The 'flame front' is the boundary between the air/gas mixture emerging from the tube and the actual flame. It is this flame front which takes on the cone shape and is the boundary of the inner cone. The cone occurs because the mixture flowing up the tube is retarded at the sides where it is in contact with the walls of the tube so it is faster towards the centre, where it reaches its top speed. This means that it tends to push the flame front away from the burner port much more at the centre than at the sides. The inner cone contains unburnt gas.

There are some burners — the fully pre-mixed burner — where all of the air required is pre-mixed with the gas usually with the aid of a fan. Fully pre-mixed burners are rarely used domestically — though they may be the burners of the future. The fully pre-mixed flame is very compact and can be incorporated into smaller appliances.

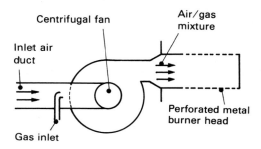

Pre-mix burner

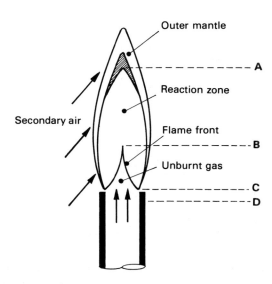

Figure 27. Zones of a pre-aerated flame. If the flame has a yellow tip, this is formed in the shaded area at the top of the reaction zone

If we look at the structure and chemistry of the partially aerated flame, assuming for simplicity that natural gas is pure methane, we see that methane and a proportion of the air required for combustion issue from the burner and approach the flame front.

Before it reaches the **primary combustion zone**, the air/gas mixture is rapidly preheated to a temperature which will sustain the chemical reaction associated with burning. The velocity of the products emerging from the primary combustion zone is greater than entry and their temperature increased.

The reason for these drastic changes stems from the fact that the reactions involved in combustion thrive on, and generate, heat.

The first stage of combustion is concentrated into a very thin 'reaction zone' less than 1 mm thick, where, because of the sudden increase in temperature on entering the zone, the reacting mixture rapidly expands in volume so that the emerging gases have a considerably increased velocity.

In most burners there is sufficient air pre-mixed with the fuel gas for complete combustion and the intense primary phase finishes when the premixed oxygen is exhausted. Air diffusing in from the surrounding atmosphere (secondary air) provides oxygen to complete the combustion process in the outer mantle. It is the combustion of the carbon monoxide and hydrogen that forms the outer mantle as a **secondary combustion zone**.

Lighting Back

If the air/gas mixture flows up the tube at the same speed at which the flame can burn it, then the flame will stay at the end of the tube, if however, the speed of the mixture is reduced, the flame will burn its way down the tube to the injector. This is called lighting back.

BS 1179: Light-back: Transfer of combustion from a burner port to a point upstream in the gas/air flow.

— Direct light back	— light back through the burner itself.
— Indirect light back Roll over	— light back by a flame not passing through the burner itself.

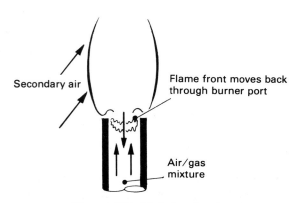

Secondary air

Flame front moves back through burner port

Air/gas mixture

Figure 28. Lighting-back

Pre-Aerated Flame: Regulation of Mixture Velocity

The size of a pre-aerated flame is determined by the quantity of air gas mixture passing along a mixing tube. The flame can be regulated to a desired shape and size by altering the mixture in any or all the following ways:

(i)   By altering the size of the gas injector.

(ii) By altering the size of the air ports.

(iii) By altering the resistance to the passage of the mixture offered by the mixing tube.

Methods of Adjustment

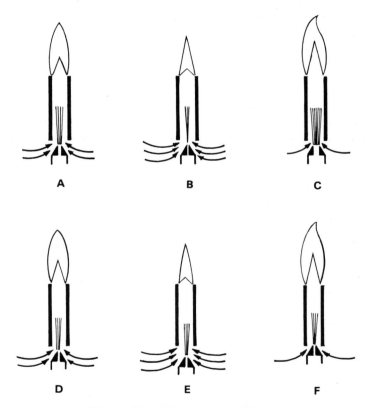

Figure 29. Adjustment of flame

A   We have a nicely adjusted aerated burner.

B   We have the same burner but the gas injector is smaller. The stream of gas being smaller has slightly less energy but has a much easier job to do in passing up the mixing tube, so that the air-gas ratio of the mixture is higher because a higher proportion of air is entrained. The flame is, therefore, sharper in character. The cone is smaller and more keenly defined.

C   We have made the injector larger. The increased quantity of gas has more work to do in propelling itself along the mixing tube and hence the air entrained is less, the air-gas ratio of the mixture lower, and the flame slacker — i.e. the cone is longer and much less well-defined. In extremes the flame would begin to become luminous.

Thus at A, B and C we have all conditions constant except the injector, and the changes have been effected by altering only the injector.

D   We have a nicely-adjusted aerated burner.

E   The air ports are larger. We have made the job of entraining air more easy and more air can be entrained. The air-gas ratio is increased and the flame becomes sharper.

F   The reverse occurs when we make the air ports smaller.

Thus at D, E and F we have all conditions constant except that we have regulated the flame's shape and size by altering the size of the air ports.

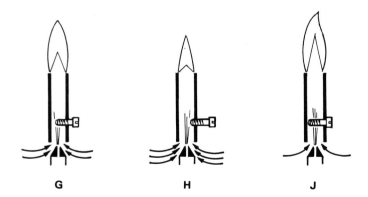

G   H   J

G   We have a properly adjusted burner. Notice also that a screw is inserted in the mixing tube of the burner.

H   The screw has been partly withdrawn. The effect is to make the job of pushing the air-gas mixture along the tube much easier. Hence there is more energy left over to suck in air. The result is that the air-gas ratio is higher and the flame is sharper.

The reverse has happened at J. By screwing the screw in farther we increase the resistance to flow in the mixing tube; more energy is used up in doing this job and less energy remains to draw in air. The flame thus becomes slacker.

Thus at G, H and J we have all conditions constant except that we have regulated the flame shape and size by inserting a screw in the mixing tube of the burner and by adjusting the screw we have altered the resistance to the passage of the mixture offered by the mixing tube.

This last method of adjustment is well worth considering carefully, since it explains the effect of any increase or decrease of the resistance in a burner brought about either by accident or design.

If, for example, dirt is allowed to collect in a burner, the flames will become slack and may even become luminous.

If, on the other hand, we reamer out the holes in a burner we shall decrease the resistance, more air will be entrained, and the flames will become much sharper. The burner may even light back.

Flame Lift

BS 1179: Separation of a flame from a burner port, whilst continuing to burn with its bore some distance from the port.

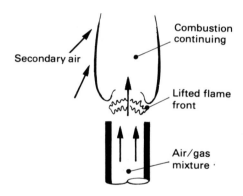

Figure 30. Flame lift

Gas Rates

All gas appliances are rated gross (input pressure) although the WATSON HOUSE plate on the appliance gives test pressures which one could assume to be nett as the test would be taken after the governor or multifunctional valve but before the burner orifice injector, this gives rise to some confusion. In real terms the input pressure to the burner is still being measured.

The Watson House Plate provides all the relevant information with regard to the particular appliance and would generally follow the format set out below:

(a)   Range Rating of Boiler or Appliance   ⎫ All data

(b)   Nominal Input of Boiler or Appliance   ⎬ in Metric and

(c)   Nominal Output of Boiler or Appliance   ⎭ Imperial units

(d)   Gas Family prepared for.

(e)   Electrical Specification.

(f)   Gas Council No. (for spares).

(g)   Manufacturers Serial No.

The calorific value of natural gas is at present 1034 Btu/ft^3 although this will vary with the purchase and blend of gas from different gas fields. It is advisable to check on your gas bill or with your local gas region should you require the exact figure at a particular time.

For calculation of gas rates and cubic foot capacities it is recognised that for every 1000 Btu of appliance/boiler rating we require 1 ft^3 of input gas (in real terms we are losing 34 Btu rating for every cubic foot of gas for every cubic foot of gas, this could lead to some discrepancy of a sizeable figure on a large unit).

A central heating boiler with a rated input of 60000 Btu (17.6 kw) would require 60 ft/hr^3 of gas input.

eg $\dfrac{60000}{1000}$ = **60 ft/hr³**

If we apply the correct calorific value

$\dfrac{60000}{1034}$ = **58 ft/hr³**

Furthermore, if we multiply the approx figure of 60 ft/hr³ by the correct calorific value we can appreciate we are really overfiring the boiler by:

60 x 1034 = **62040** Btu/hr
2040 Btu/hr overfired

The larger appliances/boilers would require the following formulae:

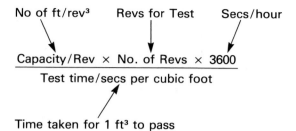

No of ft/rev³   Revs for Test   Secs/hour

$$\frac{\text{Capacity/Rev} \times \text{No. of Revs} \times 3600}{\text{Test time/secs per cubic foot}}$$

Test time/secs per cubic foot

Time taken for 1 ft³ to pass

Standard meter reading 1 ft/rev³ formulae:

$$\frac{1 \times 1 \times 3600}{\text{Test time}} = \textbf{ft}^3 \textbf{ input gas rate}$$

eg   A boiler rated at 80,000 Btu/23.5 kw input would require a gas input of:

$$\frac{3600}{\text{test time}} = \textbf{80 ft/hr}^3$$

Formulae transformed to read

$$\frac{3600}{80} = \text{test time} = \frac{3600}{80} = \textbf{45 secs for 1 ft}^3 \textbf{ gas}$$

Where the manufacturer gives a pressure set up usually in "wg or mbars we can accept this method for setting up. Although it is advisable to check with the meter throughput to ascertain the actual input rate.

mbars are equivalent to "wg × 2.5

eg   8 "wg = 20 mbars.

GAS CONTROLS

Introduction

The way in which individual control devices are arranged into a complete control system depends on the type of appliance (the way in which it is required to operate) and the type of central heating system installed.

A typical central heating system has two main areas of control:

— At or around the appliance (essential controls — primary controls).

— Remote from the appliance (ancillary controls — secondary controls).

The control system at or around the appliance is usually concerned with safety aspect in the burning of the fuel and in controlling the heat input medium used to carry the heat away from the appliance. They are almost certain to be built in as part of the appliance by the manufacturer which means that it is fairly easy to obtain precise layout, and wiring diagrams of internal components and their interconnection.

Over the years many specialised components and control systems have been devised but whatever the degree of sophistication achieved modern appliances must incorporate the following features:

— A safe and effective method or procedure for the initial lighting of the appliance.

— A flame failure safeguard to prevent a dangerous situation arising from the re-introduction of the fuel to the main burner after flame failure has occurred.

— Safety devices at the appliance as a precaution against excessive system pressures and over heating of the appliance.

— Control of the maximum heat input to the appliance so as to prevent conditions arising of bad combustion or overheating under normal conditions of usage.

— The provision of manually operated controls adjacent to the appliance for the purpose of turning off the main gas supply to the main or pilot burner.

Remote or ancillary control systems are usually connected with the attainment and maintenance of comfort conditions. Here you are largely at the mercy of the system designer in regard to the complexity of the installation and the components used. The main difficulty in fault diagnosis is to decide how the system should operate as a whole and how the failure of one auxillary component will react on the remainder of the system so that the fault may be identified.

There are many possible combinations of controls, and the examples given in this module are merely typical of the various control layouts commonly used.

72

PRIMARY CONTROL SYSTEMS

Basic Requirements

The increasing sophistication of heating systems has been paralleled by the development of appropriate controls and increasing standards of thermal comfort have given impetus to their use. Effective control of the installation is essential. Three basic objectives to which controls of domestic installations are directed are:

— Safety

— Comfort

— Economy

A central heating appliance must be capable of continuously and automatically adjusting its output to suit the needs of the customer. It must do this safely and efficiently. Many components and control systems have been devised towards this end, but whatever the degree of sophistication achieved modern appliances must incorporate the following basic features:

A safe and effective method or procedure for the initial lighting of the appliance.

A flame failure safeguard to prevent a dangerous (i.e. explosive) situation arising from re-introduction of the fuel to the main burner after flame failure has occurred.

Safety devices at the appliance as a precaution against excessive system pressures and overheating of the appliance.

Control of the maximum heat input to the appliance so as to prevent conditions arising of bad combustion or overheating under normal conditions of usage.

The provision of manually operated controls (gas cocks) adjacent to the appliance for the purpose of turning off the main gas supply to the main and/or pilot burner.

The first gas fired boilers had very basic controls, the use of automatic protection against flame failure was almost non existent. A typical boiler would rely on a substantial by-pass or a permanent pilot, a conventional gas governor and burner control cock(s). It would also have a **mechanical** fusible link arrangement providing protection against overheating.

Later developments involved control systems that incorporated flame failure protection. The first flame failure protection devices did not employ any source of electrical power but utilised either gas pressure or the heat energy of a gas flame in combination with a mechanical device or devices to secure their operation.

Pressure Operated (Non Electric) Controls

1. Basic System

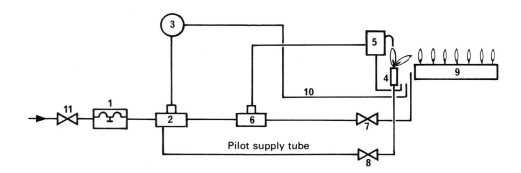

Components (for function and operation of each component see sections).

1. Main gas governor.
2. Temperature control relay valve.
3. Boiler thermostat.
4. Pilot Assembly.
5. Bi-metal cut-off valve.
6. Cut-off relay valve.
7. Burner control cock.
8. Pilot control cock.
9. Main burner.
10. Weep pipe.
11. Main gas control cock.

This basic system has the following disadvantages:

The pilot is virtually ungoverned. A governor, being designed in this arrangement to work around the full gas rate of the appliance is unlikely to successfully govern down to pilot rate (with main burner off).

By not positioning the cut-off relay valve upstream of the other components, the event of a mechanical failure of these later components or of the associated pipe-work at a time when the appliance was non operational could result in the uncontrolled escape of unburnt gas.

2. Using Pilot Governor

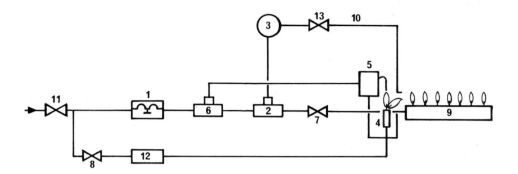

Components

1.	Main Gas Governor.	8.	Pilot Control Cock.
2.	Temperature Control Relay Valve.	9.	Main Burner.
3.	Boiler Thermostat.	10.	Weep Pipe.
4.	Pilot Assembly.	11.	Main Gas Control Cock.
5.	Bi-metal Flame Failure Device.	12.	Pilot Governor.
6.	Cut-off relay.	13.	Weep Test Cock.
7.	Burner Control Cock.		

Points of Difference

The position of the temperature control relay valve (component 2) and the cut-off relay valve (6) are transposed.

The pilot supply is taken from a point up stream of the main gas governor and is separately governed (12).

A test cock is fitted to the weep pipe (13).

Advantages

Using the pilot governor (12) helps to ensure a steady pilot-flame unaffected by changes in pressure due to valve movements of governor or relays that occur when system is in normal operation.

Using the weep test check (13) provides a convenient and easy method of checking the operation of the temperature control relay valve and/or time clock may be used to shut down boiler (except for pilot flame) by over-riding these controls.

NB Not fitted with 'Perfecta' relays.

3. Combined Temperature Control & Safety Cut-Off Relay Valve

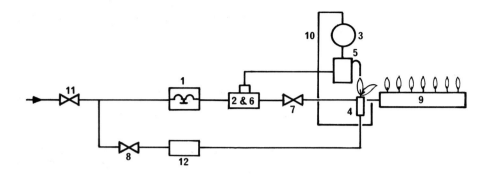

The numbering of similar components is the same as for previous layouts. The combining of components 2 & 6 is a pointer on things to come. i.e. multifunctional controls.

Points of difference

The temperature control and safety cut off relay valve are combined (2 & 6).

This layout is unusual insofar that the one relay valve is used for both temperature control and in conjunction with the bi-metal flame failure device provides flame failure protection. Because of the need for complete closure the valve for the flame failure function in this method may only be used for on-off operation of the main burner, i.e. no main burner by-pass can be allowed.

4. Thermo Electric Devices

These devices were first employed to assist in controlling the components (i.e. relay valves and other gas operated controls) and meant that gas control systems could remain much as before but with the advantage of much greater safety and reliability in flame failure protection.

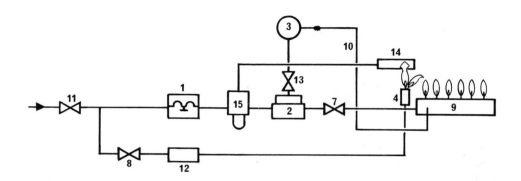

Components

1. Main gas governor.
2. Temperature control relay valve.
3. Boiler overheat control.
4. Pilot assembly.
7. Burner control cock.
8. Pilot control cock.
9. Main burner.
10. Weep pipe.
11. Main gas control cock.
12. Pilot governor.
13. Weep test cock.
14. Thermocouple.
15. Thermo electric flame failure valve.

Points of Difference

Notice that this scheme closely follows layout 2 (using pilot-governor) except that we have lost the bi-metal cut-off and one relay valve and have now introduced two new items for flame failure protection, (components 14 & 15). Note that a weep controlled relay valve is returned for overheat control. All other components function as before.

5. Unprotected Pilot Light — Flame Failure System

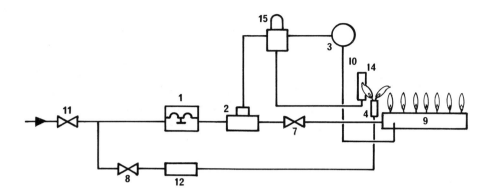

Components

1.	Main gas governor.	9.	Main Burner.
2.	Temperature Control Relay Valve.	10.	Weep Pipe.
3.	Boiler Thermostat (overheat).	11.	Main Gas Control Cock.
4.	Pilot Assembly.	12.	Pilot Governor.
7.	Burner Control Cock.	14.	Thermocouple.
8.	Pilot Control Cock.	15.	Thermo-Electric Flame Failure Valve.

Points of Difference — Indirect Control of Main Burner Gas Supply

This unprotected pilot system is similar to layout 4 except that the position of the thermo-electric F.F.D. is in the weep supply gas line. The cut-off action of the valve under pilot failure conditions is as for layout 4 but now the action is arranged to indirectly cause the temperature control relay valve to interrupt the main gas supply to the burner instead of by direct control as before.

This technique provides the same degree of protection with a smaller, lower priced component.

Note that an escape occurring on the weep line would cause a 'fail to danger' situation, in order to minimise the risk it is essential that the flame failure valve is first in position in the weep line, (i.e. directly following the temperature control valve).

N.B. The flame failure systems as shown on layouts 4 & 5 do not incorporate control of the pilot supply. Such arrangements are known as **unprotected pilot flame systems**.

6. Fully Protected Pilot System

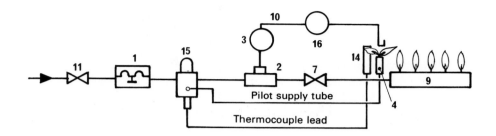

Pilot supply tube

Thermocouple lead

Components

1.	Main Gas Governor.	10.	Weep Pipe.
2.	Temperature Control Relay Valve.	11.	Main Gas Control Cock.
3.	Boiler Overheat Control.	14.	Thermocouple.
4.	Pilot Assembly.	15.	Electro-magnetic Flame Failure Valve.
7.	Burner Control Cock.	16.	Time Clock.
9.	Main Burner.		

Points of Difference — Fully Protected Pilot

In this arrangement, the flame failure valve operates as in lay-out No. 4, cutting off main gas in the event of pilot failure. The pilot supply is taken from a tapping on the casting of the flame failure valve and under this arrangement there is a 'fully protected pilot' system (i.e pilot failure means that both pilot and main gas supplies are cut off).

7. Multifunctional Controls — Non Electric

We have been looking at systems of control where all the essential control functions are performed by separate components on a gas manifold. The various arrangements of these components settled down to a fairly regular pattern and it was logical to combine the functions of the separate components into one casting.

This was the forerunner of the multi-functional controls. They seem extremely complicated but for the most part they perform the same functions as their separate relations, i.e. main and pilot gas taps, pressure governor, relay valve, thermo electric magnetic valve section of a flame failure device, pilot connection, filters and so on.

A typical non-electric control layout using this development is shown.

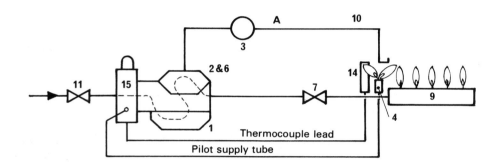

Components

1. Main Gas Governor.
2. Temperature Control Relay Valve combined in the one casing.
6. Cut-Off Relay Valve.
3. Boiler Thermostat.
4. Pilot Burner assembly.

7. Burner control cock.
9. Main burner.
10. Weep pipe.
11. Main gas control cock.
14. Thermocouple.
15. Thermo-electric F.F.D.

Points of Difference — Combining of components to form multifunctional control — components 2, 6, & 15.

Note That at (A) in the weep line a clock may be installed or even a cylinder or room thermostat, connected by extended weep line. Pilot ignition may be used on such a system by using a filament igniter head adjacent to the pilot with wires leading to a dry battery and a switch fitted to work simultaneously when the pilot button is depressed.

Electrically Operated Control Systems

It was inevitable that with the call for improved control systems that electrical devices would be introduced, which had certain definite advantages:

— Control could be applied at a point away from the appliance without the use of cumbersome extended weep lines as was necessary with non-electric control arrangements.

— Panel mounted controls were easier to arrange which helped to meet the customer's demand for a new 'clean' look.

— Electric time switches removed the necessity to wind up periodically as with the hand wound clocks.

— Electric programming was possible.

— Mains operated electrical ignition could be introduced for domestic boilers.

There is a marked disadvantage; the probability of malfunction of the system is now directly linked to the reliability of the electricity supply.

Any reluctance to use electrical equipment for gas disappeared with the introduction of small bore systems which demanded an electric supply to be brought adjacent to the boiler for the purpose of powering an electrically operated pump or accelerator, an essential component of the new system.

8. Electrically Operated Control System — Separate Components

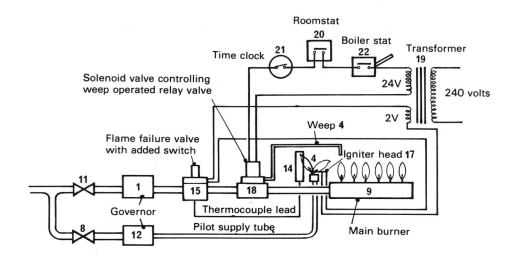

Components

It can be seen that there is a normal arrangement of gas controls but with certain additions:

15. The thermo electric flame failure valve has an independent electric switch added.

17. **Igniter:** A filament igniter is located adjacent to the pilot.

18. **Solenoid Valve:** The relay valve still retains its weep pipe but now weep gas is controlled by a solenoid valve attached to the top of the relay valve.

19. **Transformer:** A mains transformer is used to reduce the mains voltage from 250V to 12 or 24 volts for solenoid operation and to an even lower voltage, for the igniter filament.

20. **Room Thermostat:** A room thermostat being a form of temperature sensitive switch which also may interrupt the current to the solenoid valve.

21. **Time Clock (Clock Controller):** Clock control is provided but here it interrupts the current to the solenoid valve and thus indirectly controls the weep supply.

22. **A Boilerstat:** Is not really an additional component but here a different type of boilerstat is in use, i.e. this one operates an **electric** switch which, in turn, is used to interrupt the current to the solenoid valve.

MODERN CONTROL COMPONENTS

Systems Incorporating Multifunctional Controls

An 'electric' multifunctional control comprises within one casing a combination of components that are basic to both electric and non electric systems, (i.e. pressure governor, thermo electric flame failure valve, solenoid controlled gas relay valves, main and pilot gas cocks). All sections of a multifunctional control perform in a similar manner to the individual gas and electric controls already discussed.

9. Basic System

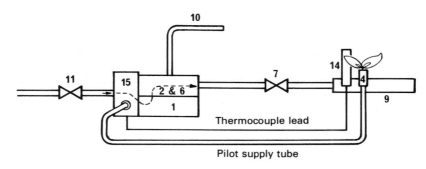

Thermocouple lead

Pilot supply tube

Components

1. Main Gas Governor.	⎫ Forms
2 & 6 Relay Valve (cut off and temperature control).	⎬ Multifunctional
15. Flame failure magnetic valve.	⎭ Control
4. Pilot Assembly.	
7. Burner Control cock.	
9. Main burner.	
10. Weep pipe.	
11. Main Gas control cock.	
14. Thermocouple and lead.	

This basic system incorporating multifunctional controls has been used in this form on many different makes of boiler. It is not complete; most of the essentials are there but no control over the relay valve weep supply and therefore no control of the main gas except by use of the main gas control cock. This layout is basic and is a very minimum, an over-heat thermostat must be added. There are many combinations of ancillary controls; room thermostats, clocks, programmers, froststats etc., these may be connected to control the main gas supply to the burner via the relay valve weep supply, either by directly interrupting the weep supply or indirectly by interrupting the electric supply to a solenoid valve or by a combination of both.

The layouts that follow show various combinations of controls that may be used to supplement the basic layout (no. 9). The multifunctional control arrangement is representative of many different makes and modules that could fulfil this function.

10. Timed control of weep using Electric Clock — overheat control by Non-Electric Boilerstat

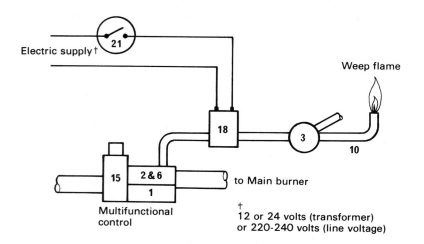

Electric supply †

21

Weep flame

18

3

10

15 | 2 & 6
1

to Main burner

Multifunctional
control

†
12 or 24 volts (transformer)
or 220-240 volts (line voltage)

Components

1. Main Gas Governor.
2 & 6 Temperature Control relay valve and cut-off relay.
15. Thermo Electric flame failure valve.
3. Boilersat — Non Electric.
10. Weep pipe to burn off point in combustion chamber.
18. Solenoid Valve.
21. Time Clock — Switch contacts only shown.

) Combined in
} Multifunctional
) Control

System Description

Overheat by non-electric boilerstat controlling valve in weep line, on/off sequence of boiler by clock contacts or programmer controlling operation of relay valve weep gas by solenoid valve.

Further Developments

The next development was the natural one of including the solenoid function with that of the Governor, relay valve temperature control relay and thermo electric flame failure in the one multifunctional control.

Thus in each of the following layouts, the solenoid valves, although shown as a separate component, may in fact be part of the multifunctional control.

NB Layouts 10/11/12 do not show the normal flame failure thermocouple and lead pilot supply shown in the basic system.

11. Overheat Control — Electric Boilerstat acting on Weep Solenoid

Components as for layout 10 but the boilerstat (Component 22) operates an electric switch which is wired in series with time clock to control weep gas via solenoid valve. (Component 18).

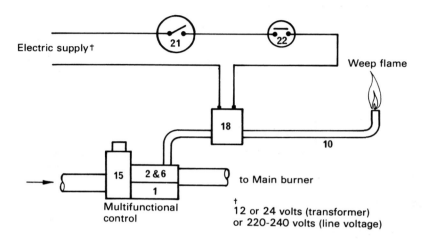

Electric supply†

Weep flame

18

10

15 | 2 & 6 | 1

to Main burner

Multifunctional control

†
12 or 24 volts (transformer)
or 220-240 volts (line voltage)

12. Roomstat Control of Accelerator Pump — Time Control of Weep Solenoid

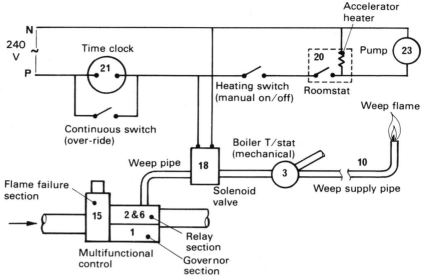

Components

1. Main Gas Governor.

2 & 6 Combined temperature control relay valve and cut off relay. } Multi-Functional Control

15. Thermo electric flame failure device.
3. Boiler overheat stat (non electric).
10. Weep pipe.
18. Solenoid valve.
21. Time Clock (Electric).
20. Room thermostat.
23. Accelerator pump.

In this circuit the relay valve weep is electrically, controlled with an electric time clock switching the solenoid.

For Boiler temperature control and to protect against the overheating of the boiler, a non-electric, rod or bellows type boiler thermostat is used in addition:

— A manually operated 'boiler on continuous' switch across the time clock is introduced to override the on/off sequencing of the clock contacts as necessary.

— There is automatic control of an accelerator pump by means of a roomstat and time clock.

— Provided there is a manually operated 'summer' switch to switch off the pump (i.e. control heating circuit) leaving the boiler to operate on gravity circulation to a domestic hot water cylinder.

13. Electric Ignition and Low Voltage Control of Solenoid

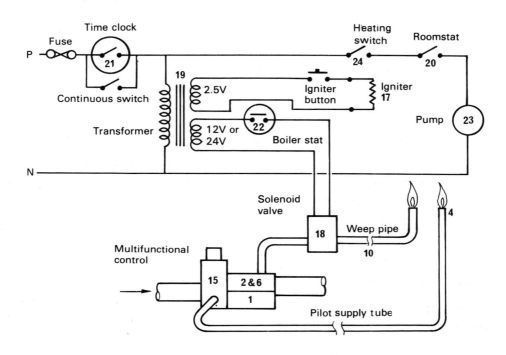

Components

1. Main Gas Governor.
2 & 6 Combined temperature control relay valve and cut off relay.
15. Thermo electric flame failure device.
4. Pilot assemply.
10. Weep pipe.
18. Solenoid valve.
17. Igniter.
19. Mains transformer.
20. Room thermostat.
21. Time Clock (Electric).
23. Accelerator pump.
24. Heating Switch.

} Multi-Functional Control

In this system there is:

— A time switch for overall boiler control (except for pilot supply). The time switch contacts operate at full line voltage.

— A transformer with:

(1) a low voltage secondary (12 or 24 volts) for the mag valve (solenoid valve) controlling the weep line gas.

(2) a low voltage secondary (1.5 to 6 volts) for an igniter which has a press button operated switch.

— An electrical boilerstat to interrupt the weep solenoid current as both means of controlling boiler temperature and as a precaution against overheating of the boiler.

— An accelerator pump — mains operated, to propel water around the central heating circuit.

— A line voltage roomstat controlling the pump.

— A manually operated 'Heating Switch' which allows the heating circuit to be put out of action by stopping the pump, leaving the boiler to operate under the control of the clock and/or the boilerstat for the gravity heating of domestic hot water.

14. Low Voltage Ignition/Mains Voltage Powered Solenoid Valve/Roomstat Control of accelerator Pump

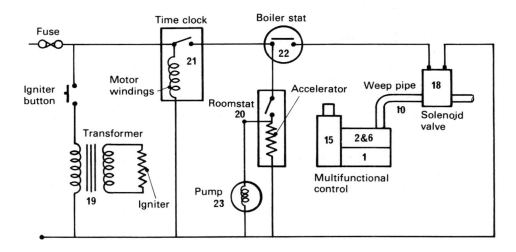

Ignition:	Transformer operated low voltage igniter. Ignition 'on' switch in transformer primary circuit.
Boiler Control:	Electrical Boilerstat for boiler temperature control and overheat protection in conjunction with solenoid valve acting on relay weep line. Overall on/off operation of boiler by electric time clock.
Space Heating Control:	A roomstat with parallel accelerator, operates at line voltage. Controls accelerator pump on heating circuits, (i.e. pump off — no circulation of water to radiators but boiler remains on under the control of the boilerstat).

Multifunctional Gas Control Valve, Operational Sequence

1. To light the gas appliance the reset button should be fully depressed, this allows gas to flow through to the pilot, but at the same time closes the flow interrupter valve thus closing off any flow of gas through the main body of the valve, and also presses the thermocouple armature against the magnet assembly.

2. If the pilot is lit it will heat a thermocouple mounted adjacent to the pilot, after a period of time (max. 40 secs) the magnetic field will be of sufficient intensity to hold the armature against the pull of the operating spring attached to the magnet assembly.

3. At this point the reset button can be released, this opens the flow interrupter valve to establish a gas flow to the burner.

4. The required pressure is set by adjusting the remote regulator which in turn determines the position of the main pressure diaphragm.

5. Should the pilot be extinguished the thermocouple cools thus causing a failure of the magnetic field allowing the operating spring to slam shut the interrupter valve and the pilot seat. The valve is now at a new light up position involving a repetition of the operating sequence.

6. The electrical control side of the valve may be 24V ac or 240V ac and consists of a solenoid valve which when energised operates a control valve allowing remote regulator pressure (4) to pass to the main diaphragm. When the heating load is satisfied the solenoid is de-energised closing the control valve cutting off the remote regulating pressure to the diaphragm rendering the diaphragm closed thus shutting off the main gas to the appliance.

Main Gas Control: Solenoid valve (section of multifunctional control) controls weep gas of weep operated relay valve. Remainder of multifunctional control comprises governor, main and pilot cocks, flame failure valve, pilot off-take.

15. Direct Control of Main Burner Gas using Solenoid Valve

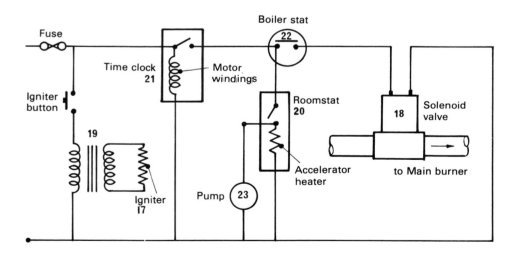

Domestic hot water is by gravity. Control is by Boilerstat and/or clock control only.

Time Control: Time clock arranged for overall control of boiler.

Ignition: Transformer operated low voltage ignition.

Boiler Control: Boilerstat with electric switch controlling solenoid valve circuit.

90

Main Gas Control: Solenoid valve **directly** controlling main burner gas.

Space Heating Control: Roomstat with a parallel accelerator, operating at line voltage. Controls accelerator pump on heating circuit.

Other Control Systems: Although the basic control arrangement for domestic boilers has settled down as regards the essential functions of flame failure, safe means of ignition and overheat prevention there is no limit to the number of the ancillary components that might be used with a particular system.

Modern Multifunctional Controls

We have mentioned the progress towards the modern multifunctional control which, amongst other features dispenses with the need for cumbersome weep lines. Modern systems are 'all electric' insofar that programmers, room thermostats, overheat stats etc all serve to control the electric supply to the multifunctionals 'built in' solenoid which in turn controls the gas supply to the main burner as required.